The Pomegranate

by :

S. S. HIWALE

Principal Scientist

Central Horticultural Experiment Station (C.I.A.H.-I.C.A.R)

Godhra-Vododara Highway, Vejalpur - 389340

Gujarat, India

2009

New India Publishing Agency

Pitam Pura, New Delhi- 110 088

Published by
Sumit Pal Jain *for*

New India Publishing Agency

101, Vikas Surya Plaza, CU Block, L.S.C. Mkt.,
Pitam Pura, New Delhi- 110 088, (India)
Phone: 011-27341717, Fax: 011-27341616
E-mail: newindiapublishingagency@gmail.com
Web: www.bookfactoryindia.com

ISBN : 978-9380235-15-8

Typeset at: Typographiya # 98 11 48 23 28
Printed at: Jai Bharat Printing Press, Delhi

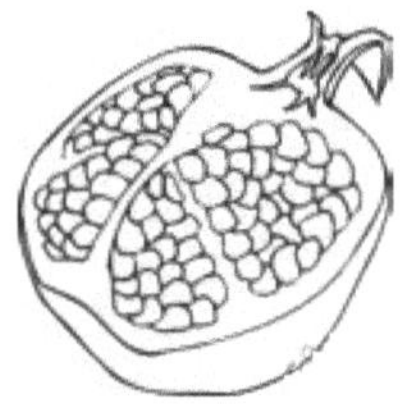

The Pomegranate

Foreword

Pomegranate is one of the important semiarid fruits cultivated in over 1.25 Lakh hectare area, commercially in Western India. Anardana type is cultivated in H.P. and J.K. It is highly specific to climatic requirement and its cultivation is restricted to particular climatic region. Keeping in view the new economic policies and globalization, the Pomegranate growers will be challenged to increase the productivity to meet national as well as international standards to meet internal and export demand.

The book "Pomegranate" is one of the most comprehensive treatise on collection and conservation of germplasm, production and postharvest management of the crop. The compiled information in the book will cater to the needs of Researchers, Teachers, Policy makers, Agricultural officers, Students and Pomegranate growers. The scattered information published on pomegranate, in various journals, books, bulletins *etc.* are elaborately covered in this book .The chapters are intended to present the state of the art of technology advancement, which has taken place over the time.

I am confidant; this book will be of tremendous help to all involved with pomegranate in formulating research strategies besides harnessing the untapped opportunities available in pomegranate cultivation.

Sd.

M. C. Varshneya
Vice Chancellor
Anand Agricultural University
Anand, Gujarat

Preface

Pomegranate (*Punica granatum* L) is emerging as one of the important fruits of semiarid rainfed areas. Its cultivation is possible even on marginal degraded lands earlier found unsuitable for growing crops. Apart from this its ability to with stand salinity in soil and water to some extent made this crop to emerge as a hardy fruit crop. Research carried out over the years at Ganesh Khind Research Station at Pune in Maharashtra by Dr. Cheema gave boost to put more area under this crop. In the recent times it is the leading fruit in Maharashtra with an estimated area of more than 1.25 Lakh hectares and its increasing importance as major export earner. Earlier the requirement of Pomegranate and Pomegranate products in our country had been met through import, which put a heavy burden on the exchequer for foreign exchange. This prompted the Government of India to launch the projects on the improvement of Pomegranate and Pomegranate culture in India. During the last 3 decades enormous improvement has been made in this field with the evolution or selection of commercial cultivars suitable for North, South, and West Indian regions and standardization of their agro-techniques and post-harvest technologies. However, only a fraction of information has passed to the growers and research organizations due to transmission gap. Even this meagre information has improved cultivation so much that it has not only lessened the burden on foreign exchange by minimizing the import of Pomegranate, but also the exchequer is now being augmented partly through the export of Pomegranate.

The book contains work carried out by various workers in the past 30 years in India; on various aspects. The book contains chapters on Species, Uses, Composition, Plant genetic resources and their evaluation, Breeding, Varietal improvement, Commercial varieties. Cultural management of the crop includes climate and soil requirement, Propagation and nursery management including cutting, air layering, Micro propagation and Economics. Bahar treatment, flowering, fruiting, Irrigation, Drip and fertigation, Manures and fertilizers including biofertilizers, organic fertilizers. Apart from this aspects like High density orcharding, Canopy management, Fruit thinning, chemical regulation, Staggering of crop, Rejuvenation of old orchards, Pest and diseases management, Postharvest technology and value addition, Postharvest handling for export, Standards for pesticide residue, Glossary of terms, Literature cited *etc*. are all covered in the book with around 100 pages and 22 photographs.

Author is also grateful to the Director, IIHR, Bangalore, CIAH, Bikaner and Head CHES, Vejalpur for encouragement, technical and scientific guidance. Maharashtra pomegranate grower's organization for giving opportunity to interact with researchers as well as growers during their yearly seminars. Thanks are due to Shri R. B. Baria, Technical assistant for typing the manuscript and Shri C. S. Chamar, Technical assistant for evaluating the crop and maintaining the germplasm.

S.S. Hiwale

Gujarat

Contents

Chapter 1

INTRODUCTION

Introduction

Pomegranate is emerging as one of the important fruit crops owing to its hardiness and ability to withstand adverse soil and climatic conditions. Today out of world production of ten Lakh tons; India contributes 40-45 percent. Currently the export of pomegranate is around 10,000 tones, which is just 5 percent of the total export. With the increase in population the domestic demand of fruits has also increased substantially. As on date out of total area of 1,25,000 ha. area, under pomegranate major area under the crop is found in the state of Maharashtra. The state has 85,000 ha. (68 % area) under the crop, of which only 51,000 ha is under production, followed by Karnataka 12,727 ha. Gujarat 3,787 ha. up to 2003 -2004. Earlier orchards were raised mainly with variety Ganesh, which was the leading variety. However the variety is fast being replaced with the new cultivars like Bhagawa and Arakta, which have attractive skin and aril color and hence are suitable for export.

The pomegranate is an ancient and favorite table fruit of tropical countries. It is commercially grown for its sweet-acidic fruits, which provide a cool refreshing juice, and is valued for its medicinal properties. The fruit rind, juice, leaf and roots are used in the preparation of various ayurvedic medicines. Its popularity is also due to the ornamental nature of the plant especially due to bearing bright red flowers throughout the year. The fruit juice is refreshing and contains both glucose and fructose. The juice and seed contain large quantities of tannin and agolic acid, which is essential in curing several diseases. The fruit can be successfully grown even under purely rain fed conditions. In the recent years pomegranate cultivation has became an economically viable preposition.

Species

Pomegranate (*Punica granatum* L.) belongs to the natural order Punicaceae; Punica perhaps is the only known genus of this family, which includes large shrubs or small trees with 2 species, *Punica protopunica* Balf. is found wild in Socotra Island and the other *Punica granatum* cultivated in the sub-tropical parts of the world. *Punica granatum* has been classified into two sub-species chlorocarpa and porphyrocarpa, each having two varieties. These sub-species have been established on the basis of the colour of the ovary, a stable feature that is retained even when they are reproduced by seeds. Sub-species chlorocarpa is mainly found in the *Transcaucasus*, whereas, the second Sub-species porphyrocarpa is mainly distributed in central Asia.

Uses

The pomegranate is a favorite table fruit in tropical countries. It is very much liked for its cool refreshing juice and also valued for its medicinal properties. The juice is considered to be useful for patients suffering from leprosy. The bark and rind of the fruits are commonly used in dysentery and diarrhoea. The rind is also used as dyeing material for cloth. Fresh juice of pomegranate retains its flavor and keeps for a long time when bottled after sweetening with sugar. A kind of wine is also prepared from pomegranate juice and this wine is considered superior to grape wine.

Composition

Water	78.2%	Potassium/100g	133 mg
Protein	1.6%	Iron/100g	003 mg
Fat	0.1%	Thiamine/100g	0.6 mg
Fibers	5.1 %	Riboflavin/100g	0.1 mg
Calcium/100g	10 mg	Nicotic acid/100g	0.3 mg
Magnesium/100g	12 mg	Vitamin C/100g	14 mg
Phosphorus/100g	70 mg	Oxalic acid/100g	14 mg

Now a days importance of Pomegranate in countries like USA and other European countries has increased due to its medicinal value due to which it is included in 32 out of 56 groups of pharmacological and therapeutic treatments. According to one study anthocyanins in the pericarp have antineuplastic effect. Polyphenols were found of suppress growth of cancer cells. Coumarins have hypertensive, anti-coagulant, analgesic bactericidal and anti-tumor activities-oxy commarins can prevent clot formations and blood vessel nupture Riboflavin (vitamin B_2) nonmalizes nervous system also useful in treating radiation sickness gallic acid is an antisclerahl compound reduces the level of cholesterol in blood). Tannins a alkoloid can be used to treat gastro intentinal disturbances. Seed of pomegranate contain estrogens. Its antioxident property can be exploited to counter act tree radicals and pomegranate is ranked as number one antioxident.

Chapter 2

VARIETAL IMPROVEMENT

Breeding

Pomegranate is genetically heterozygous material thus there is lot of variability existent in the nature. Therefore there is scope for crossing two selected genotypes. Raising of hybrid population and identification of superior types for following major constraints in its successful cultivation.

1. Cracking and splitting of fruits is a severe menace due to variation in atmosphere, humidity.
2. Resistance to fruit borer and stem borer.
3. To develop varieties having high sugar (16%) and less acid (1.5 %).
4. All varieties bear some sterile flowers. This reduces the yield appreciably. Selection of varieties bearing high percentage of fertile flowers should be achieved.

5. To develop better varieties for Western India by hybridization between the locally adapted *Alandi* and *Dholka* types. *Alandi* types are having medium sized fruits, deep red to deep pink flesh, hard seeds and quite sweet juice but are slightly more acidic than *Dholka* types. *Dholka* types bear larger fruits with greenish white rind, whitish flesh, soft seeds and sweet juice.
6. To develop better varieties for Southern India by hybridization between prolific bearing Paper Shell or Spanish Ruby, and the soft seeded Musket Red and the large-sized-fruit bearing Vellodu types.
7. Breeding varieties with dark red aril and skin color.
8. Breeding anardana type varieties with dark red and bold arils, high acidity, bigger fruit size, high yield.

Flowering Behavior

1. Inflorescence is either dichasial cyme or the flower are terminal and solitary.
2. Flowers are complete, actinomorphic, bisexual and epigynous.
3. Calyx is five to seven lobed, fleshy, yellowish-red colored, campanulate or bell-shaped and adnate to the ovary. The aestivation is valvate.
4. Corolla consists of five to seven lustrous red but wrinkled petals. These are crumbled in bud emerging from the edge of the hypanthium (Calyx tube). The aestivation is imbricate.
5. Androecium consists of indefinite stamens emerging in several whorls from within lower half or more of hypanthium (Calyx tube). Filaments are slender, free and equal in length.
6. Anthers are bicelled, dorsifixed and dehisce longitudinally.

7. Gynoecium comprises one to twelve carpels, but many species are tricarpellary. These are syncarpous.
8. Ovary is inferior and is embedded in thalamus. Style is yellow and fleshy with greenish stigma.
9. Anthesis commences from 8 a.m. to 1 p.m. varying with the variety. Opening is influenced by temperature, and is hastened by higher temperature. Dehiscence commences from 8 a.m., and continues till 3 p.m., varying with the variety.
10. Stigma becomes receptive prior to anthesis and remains so for two days after anthesis.

Breeding Methods

Selfing

Bag the bud a day prior to anthesis. Put a tag and label it.

Crossing

Select appropriate young bud, which will open the next morning. Insert forceps through petals or remove one petal and take out the stamens. Take care that no anther is left inside the flower, and the ovary is not injured. This can be done easily as stamens emerge from within upper half or more of hypanthium (Calyx tube) and are distant from ovary. Bagging and tagging is done. Collect the desired anthers in yellow stage and incise them with needle and transfer the pollen with the help of a brush on to the receptive stigma of the emasculated bud, the next morning. Bagging and labeling is done.

Hybridization

Hybridization is done in the morning hours before 11a.m. The steps involved are emasculation (removing of anthers) by using forcep. Pollination with the pollen of desired male parent. Covering with butter paper bag and labeling it Fruits require about 120-130 days for maturity. Crossed fruits are harvested and the seeds extracted. They are sown in plastic trays. Seeds germinate with in 10 days of sowing. The seedlings are transplanted to polythene bags and subsequently to field for further evaluation, Bhandari (1979).

Types and Varieties

Several types of pomegranates with varying shapes, sizes and color are cultivated in India. Fruits are round, obvate in shape and vary in diameter from 8-12 cm. The rind may be thick or thin and the color ranges from pale yellow to crimson. The seed pulp in superior types is thick, fleshy and very juicy, while in inferior types it is thin. The seed coat varies in hardness, some of the softer seeded types being known as seedless (Bedana). Lack of lignification of the testa is the main cause of so called seedlessness in pomegranate. Among the numerous types grown, the Bedana and Kandhari are considered the best.

The varieties Ganesh, Mridula, Arakta, G-123, P-23, P-26, Mastani, and Bhagwa are popular in Maharashtra. Bassein Seedless, Jyothi, Ganesh, Mridula, Arakta and Ruby in Karnataka, Dholka in Gujarat, Jalore Seedless, Jodhpur Red and Jodhpur White in Rajasthan, Kabul Red, Vellodu, Yercaud-1, CO-1 in Tamil Nadu.

Though we know a great deal about pomegranate and their origin there are a number of seedling varieties in the pomegranate growing areas. Selecting a variety with known qualities is always the better choice. Most horticulturists divide pomegranate varieties into three categories Sweet, Sweet tart and Sour, we prefer to include the seed hardiness in this and divide them into six groups.

1. Soft seeded sweet
2. Soft seed sweet tart
3. Early variety (mostly sweet)
4. Normal (harder) seeded sweet tart
5. Normal (harder) seeded sweet
6. Sour (Nearly always normal seeded)

The pomegranate varieties ruling in different countries are:

Spain

1. *Molar* : The best seedling pomegranate in Spain, sweet good red fruit, productive
2. *Valencia* : An early ripening sweet variety with moderate production

Malta

1. *Blanca*: Medium whitish fruit, seeds are undeveloped (nearly seedless), aril pink in color, productive.
2. *Dulce Colorada*: Large red color, red arils, soft seeded, sweet, productive.
3. *Gorda de Jativa*: Large red fruit, red aril, sweet, productive.
4. *Pignoncea*: Large red fruit, small seed, sweet.
5. *Cagin*: Large red fruit, very hard small seed, sour.
6. *Quiseppe*: Large, red, sweet fruits with high production.

Italy

Dolce, Alappi, Dolce Nostrale

Cyprus

Zaitiki : Large, sweet fruit with soft seed.

Greece

1. *Patras acide* : Extra large, red fruit with very sour taste, good for syrup
2. *Patras Douce* : large red, sweet fruit, productive

Yogosalavia

Slatki Barshi Nar : Early ripening, large, yellowish green fruit with sweet taste

Bulgaria

1. Uzbekskii Saldikii (Uzbek sweet)- sweet fruit
2. Nikiskii Rannii - Sweet tart fruit, early ripening and early to produce

Israel

1. *Wonderful* : The sweet tart variety from USA. High production with medium hard seeds.
2. *Asmar (Black)* : Dark purple almost black fruit, white, hard seeds
3. *Ras-el-Bghil* : Very large fruits with normal seed
4. *Red Lufani* : Red, large with normal seed, good production.
5. *Malissi* : Nearly seedless (called seedless) early ripening.

Arabian Peninsula

1. *Roman chouall* : Medium dark, almost dark fruit, sweet tart flavor, grown in Iraq.
2. *Mellassi* : Large red fruit with soft seeds and sweet taste.
3. *Cherabani* : Medium red fruit, sour taste and the syrup is used for preparing beverages.
4. *Selimi* : Large red fruit with sweet tart taste and small seed.
5. Aswar (Black), Ahmar (Red) and Halwa also grown.

Turkey

1. *Hicaznar* : A large fruited, having good productivity.
2. *AK-anar* : Yellowish white juice & light seeds.
3. *Kizil anar* : Large red fruit with sweet tart flavor, high production.

Iran

Shahvar (Saveh)- Red skin, very large fruit.

Afganistan

1. *Kandhar large white* : Extra large yellowish sweet fruit, early ripening.
2. *Kandhar large red* - Extra large fruit with sweet flavor, mid season ripening. This is one of the best variety in the world of Pomegranate.
3. *Kandhar large black* - Dark purple (looks black), extra large fruit, sweet tart fruit.

India

1. *Ganesh* - The number one pomegranate in India, large yellowish red, with soft seeds.
2. *Mridula* - A hybrid of Ganesh and a red Russian variety, pink red skin with soft seed, sweet taste, early mature in 150 days.
3. *Bhagwa* - Sweet, soft seeded, red fruit suitable for export.
4. *Muscat* - The rind is yellowish pink, grains are medium hard.
5. *Bassein seedless* - The outer skin color is red, seeds soft.
6. *Jodhpur Red* - Grown in Rajasthan, rind is yellowish in color with whitish pink arils.
7. *Jalore seedless* - Grown in Rajasthan the rind is reddish. Arils are Soft, pink colored.
8. *Dholka* - Grown in Gujarat, arils pink red colored.
9. *Joyti* - Released from UAS, Bangalore, red fruit with soft arils.
10. *RCR-1* - Released from UAS, Dharwad. Fruit red coloured soft arils.

11. *G-137* - Open pollinated variety from Ganesh, released by MPKV, Rahuri.
12. *Yercaud Selection: CO-1*- Released by TNAU, Coimbtore having soft seeds.
13. *Ruby red* – It is a multiple hybrid developed at IIHR, Bangalore, rind red, seeds soft. Fruit size small.

China

1. *Cin – Kwene-liu* - Cin –pehin e-liu are some of the varieties.
2. *Echen* – Large thin skinned (paper shell) fruit, sweet taste.

Tunisia

1. *Gabsi* – Large yellowish fruit, with sweet taste.
2. *Tunsi* – Medium red fruit, dark red aril with sweet tart taste.
3. *Chelfi* – Medium red, seedless variety.
4. *Djebeli* - Very large red fruit, late ripening sweet variety with very small seed.

Mexico

Granada de china and Granada Agria are grown.

United States of America

1. *Granada* – Very early, red medium a bud sprout of wonderful.
2. *Early foothill* – Large fruit similar to wonderful, ripens after Granada, low acid and high sugar.
3. *Early wonderful* – Large, early.
4. *Wonderful* – Bright red variety with sweet tart. This is the main commercial variety of USA. Very productive in San Jaquin Valley of California. It is frost sensitive variety.

Arka Ruby

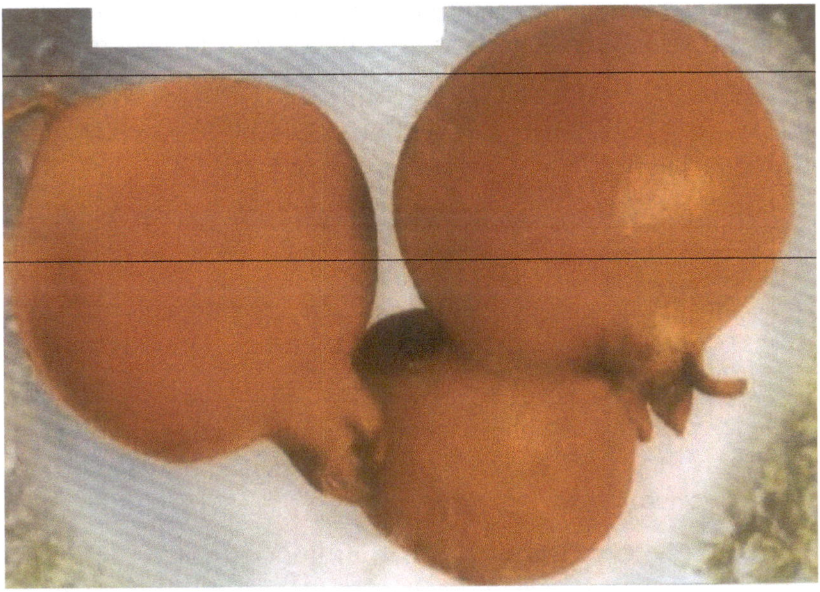

Bhagawa

Commercial Varieties

Bhagawa : The variety is known by various names viz. Shendri, Asthagandha, Mastani, Jai Maharashtra, and red Diana in various parts of Maharashtra. The variety is as on date is the leading variety of Maharashtra state. It is a heavy yielder and posses desirable fruit character like attractive fruit colour, dark red arils, no cracking and hence found suitable for domestic as well as export market. The fruits mature late in 180-190 days after fruit set. Average yield is 30-40 kg fruits / tree. The fruit is having attractive saffron colored thick, glossy skin, arils dark red and bold. The variety is less susceptible to fruit cracking, fruit spot and thrips. The market price is 2-3 times higher than other varieties.

Phule Arakta : The variety is released by Mahatma Phule Krishi Vidyapeeth, Rahuri MS. It is a heavy yielder and suitable for export due to attractive skin and aril color. Fruits are ready for harvest within 120-130 days of fruit set

Alandi : Spreading type variety, produces medium sized fruits with hard seeds, arils are red in color and sweet in taste.

Ganesh : It is selection from musket. Large area in Maharashtra still under this variety. Suitable for export, soft seeded, arils pink in colour and sweet in taste.

Dholka : Medium yielder, bears large fruits (400-500 g), rind greenish white, Fleshy testa and juice acidic in nature.

G-137 : It is a clonal selection, producing pink colour fruits. Juice content is high 70-80 percent, bold arils. medium sized fruits.

Jalore seedless : It is an early ripening variety known in Rajasthan. It produces small size fruits with soft seeds. It is tolerant to fruit cracking.

Jyoti : It is promising type, earliar known, as GKVK-1. It is a selection from basin seedless. It produces fruits with soft seeds and red aril color.

Ruby : It is a composite hybrid. It is released from Indian institute of Horticulture research, Bangalore. The fruit size is small having deep red arils. The variety is a medium yielder.

Amlidana : Is a F_1 hybrid (Ganesh x Nana). It grows well under tropical climate with quality fruit attributes. Amlidana is superior to sour variety Daru, whose trees come up naturally in temperate region of north India. Its fruits provide highly acidic anardana (16-18 %) and higher fruit yield/tree in addition, its short statured trees are suitable for High-density planting, giving increased fruit yield/unit area and hence recommended for commercial cultivation (Jalikop *et al.*, 2002).

Chapter 3

PLANT GENETIC RESOURCES AND ITS EVALUATION

The world's largest collection comprising 1117 entries is maintained at Turkmenistan (CIS). They exhibit great variability for fruit size and quality, skin color, resistance to abiotic and biotic stresses but most of them are temperate in nature and do not grow well in the Indian tropics. There is scope to use this material for breeding superior types.

There are several varieties adopted to both tropical and subtemperate climate. They are either evergreen on semi deciduous or deciduous types. The flowering habit, fruiting and flower physiology are altered with the habitat. The tropical climate with mild winter of south India, growth and flowering are continuous processes, while in the subtropical climate of north India the trees remain dormant during cold winter and flowers in the following spring and in temperate climate flowering occurs in summer.

In India cv. 'Ganesh' was one of the leading variety. It is a seedling selection from 'Alandi' having large fruit size of 400-450 g with sweet (16-17°Brix TSS) and soft seeds but the color of the aril is not attractive (pink or light pink), later on two new varieties named Arkta from MPKV. Rahuri and Ruby from IIHR Bangalore having red aril color were released using Ganesh as a base. In both the varieties red aril color was imparted by Russian varieties Gulsha red and Gulsha Rose Pink. Ruby is a multiple hybrid resembling more of Ganesh while Arkta is a F2 selection. Recently a new variety named Bhagwa has been planted in a large area in Maharashtra, which is selection made by a farmer. It had deep red aril color and attractive skin color but the variety matures late (160- 170 days). There are several other seedling selections, which are grown on a limited scale across the country, they are Madugiri, G-137, Muscat, Jyothi, Bassein seedless, Dholka, and Jalore Seedless in the states of Karnataka, Tamil Nadu, Haryana, Gujarat and Rajasthan. In Himachal Pradesh a sour type known as Daru is found abundantly in wild, is mostly used for Anardana preparation. A new F1 hybrid developed at IIHR Bangalore named "Amlidana" has been released. Most of the varieties have one or the other superior trait but are susceptible to Bacterial nodal blight, a disease that has threatened the pomegranate cultivation. The quality parameters are highly influenced by the soil and climatic conditions. There is a need to develop a variety with desirable traits like bold arils, red color with soft seeds, resistant to blight, drought, cracking, and aril browning.

Evaluation of pomegranate germplasm under semiarid rainfed conditions indicated that seeded varieties bear fruits with higher fruit weight, where as their aril seed ratio was the least and 100 seed weight, the highest. Seasonal variation affected the physico-chemical characters studied. In Hasta bahar the quality of the fruit was the best compared to Mrig bahar fruits. All the characters showed increasing trend except acidity, which declined in Hasta bahar. The seeds of hasta bahar were harder than mrig bahar fruits but the juice color was attractive in hasta bahar. Fruit weight in Mrig bahar ranged from 98 g to 200 g whereas; in hasta bahar it ranged from 141 to 261 g. Fruit length varied from 75.5 to 50.0 mm and 75.5 to 66.5 mm in Mrig and Hasta bahar respectively. Fruit diameter varied from 50.05 to 73.5 and 7.02 to 78.0 in Mrig and hasta bahar respectively. Aril seed ratio varied from 0.63 to 7.42 in Mrig and 0.69 to 3.02 in Hasta bahar. Juice percent varied from 58.3 to 77.3 in Mrig bahar and 62.29 to 78.25 in Hasta bahar. Acidity of the fruit ranged from 0.35 to 1.45 in Mrig and 0.28 to 0.90 in Hasta bahar, Hiwale (1998). Purohit (1985) in his studies on seed softness of commercial varieties reported that thickness of the testa, weight of whole seed, density of the seed, and weight of testa as percent of seed weight can be used to determine the softness of seed of pomegranate. The scale fixed by him indicated that variety having thickness of seed testa <0.5mm, the density of whole seed< 0.6g/ml, the density of seed testa <0.4g/ml, weight of the seed testa <50 percent of the weight of whole seed. Bedana, Bassein seedless, Ganesh and Dholka were found to be soft seeded. Though Bedana had soft seeds it cannot be grown commercially owing to its deciduous nature and sparce flowering.

Pomegranate germplasm evaluation of 24 genotypes carried out by Meena *et al.* (2003) indicated that all the characters except total soluble solids and total sugar had less pronounced differences between the value of phenotypic and genotypic coefficient of variation. Genotypic coefficients of variation were higher for seed content, wt. of 100 aril, wt. of 100 seeds, acidity, average fruit weight. The results indicated that higher magnitude of variation for above traits offer better opportunities for improvement through selection. These results are in consonance with those of Chaturvedi *et al.* (1980) and Pandey and Bist (1998). Higher estimates of heritability were obtained for all the characters except total soluble solids, ascorbic acid and total sugar content, which may be due to environmental effect on these traits. Acidity and wt. of 100 seeds had high heritability in spite of low genetic advance, which may be due to introduction of replication in the system (Burton and Devmore, 1953). While low genetic variability and low genetic advance was noted in total soluble solid. High heritability estimates coupled with high-expected genetic advance as percent of mean were observed for mean seed content, wt. of 100 seeds, acidity content and average fruit weight indicating that high heritability coupled with high genetic advance might be due to additive genes effect on these traits. These characters can be considered as reliable selection indices. High to moderate heritability along with low genetic advance were recorded for total sugar and ascorbic acid content indicated that these characters might be controlled by nonadditive genes action and improvement through selection for these characters would be rather ineffective. Panse and Sukhatme (1978) expressed that if a character is governed by additive genes action, heritability and genetic advance both would be high. High estimates of heritability along with high genetic advance provide good scope for further improvement in advance generation, if these characters subjected to mass progeny or family selection.

Anardana type of germplasm lines collected at CHES, Vejalpur were evaluated under rainfed conditions and all the lines flowered and fruited in all the bahars. Some lines had the desirable characters like dark red arils, soft seeds, high acidity, bold arils, high sugar acid ratio etc. Particularly line H had high acidity, big fruit size with soft seeds the aril color was light pink, Hiwale (2001).

contd...

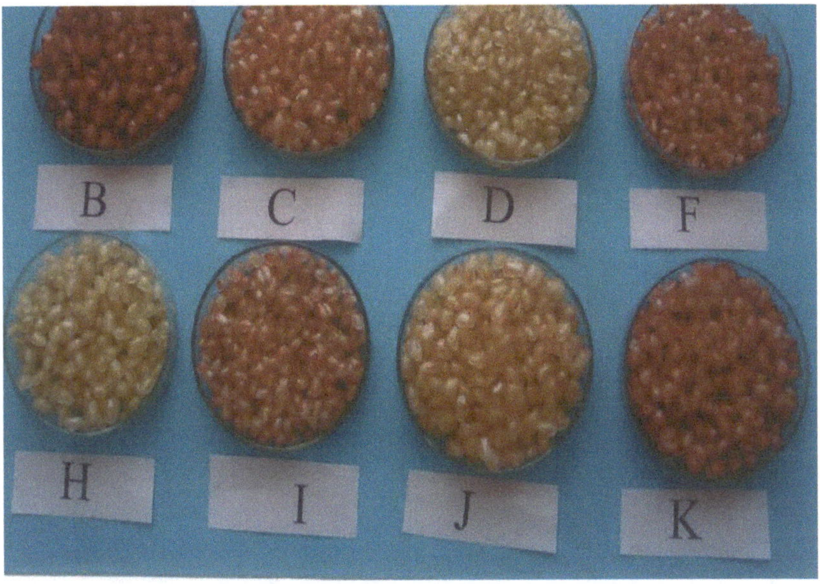

Pomegranate Anardana Type

Anardana

Germplasm evaluation in pomegranate

Variety	Season	Fruit Wt. (g)	Fruit Length (mm)	Fruit Dia (mm)	Aril /seed ratio	Juice %	T.S.S. ° Brix.	Acidity %	Juice Color	100 seed Wt.(g)
P-16	Mrig	110.48	61.5	57.5	1.26	67.17	13.3	0.462	Cream	1.96
	Hasta	141.83	70.9	70.24	3.03	78.00	16.6	0.369	Light pink	2.57
P-26	Mrig	153.32	69.1	65.5	1.41	75.21	14.0	0.351	Light pink	2.14
	Hasta	233.00	74.5	75.5	2.19	73.15	15.4	0.31		2.57
Kabul	Mrig	105.15	62.0	59.5	0.88	71.28	13.4	0.427	Light pink	1.63
	Hasta	194.00	74.5	73.5	1.30	75.43	16.4	0.28	Red	2.12
Jalore	Mrig	095.73	54.75	50.5	0.93	61.70	12.4	0.78	Cream	1.87
seedless	Hasta	176.50	67.5	70.5	1.35	76.14	15.2	0.28	Pink	2.25
Musket	Mrig	198.20	75.5	73.5	1.42	74.06	14.4	0.46	Cream	1.55
	Hasta	227.00	71.5	76.2	1.57	77.17	16.4	0.28	Pink	2.49
Jyoti	Mrig	192.14	71.4	73.0	1.41	73.69	14.4	0.35	Light pink	1.29
	Hasta	215.00	73.5	75.5	1.64	78.25	18.0	0.35	Pink	1.85
Ramna-	Mrig	200.68	71.0	70.5	0.63	58.32	9.4	0.63	Light pink	3.76
aram	Hasta	261.00	75.5	78.0	0.69	64.92	13.4	0.49	Red	4.77
Gulsha	Mrig	127.62	56.0	64.5	0.70	77.30	11.4	1.45	Red	2.98
Rose	Hasta	188.00	66.5	75.0	0.90	73.75	13.6	0.90	Dark red	3.56
Appuli	Mrig	117.38	50.05	53.5	0.71	67.05	13.4	0.63	Light pink	2.89
	Hasta	216.00	67.5	74.0	1.34	62.29	16.2	0.49	Pink	3.66
Ganesh	Mrig	134.00	71.3	70.1	1.89	77.0	16.6	0.44	Light pink	1.51
	Hasta	141.00	72.5	70.2	2.70	77.0	17.6	0.34	Pink	1.96

Variation in physico chemical characters in Anardana type pomegranate germplasm

Line	Fruit set/ plant	Fruit wt (g)	Fruit length mm	Fruit Dia. mm	Skin wt (g)	Total aril wt (g)	Juice %	TSS °Brix	100 Aril wt (g)	Aril Length mm	Aril Dia. mm	Juice color
A.	85.0	77.66	68.3	58.9	53.3	95.33	35.2	17.1	95.3	9.76	6.5	Light Pink
B.	59.0	67.00	54.0	48.4	22.0	43.66	39.4	11.1	42.3	9.46	5.16	Pink
C.	17.7	104.0	64.5	61.3	37.0	80.66	37.0	17.0	83.6	9.76	5.03	Dark Pink
E.	48.7	86.33	50.5	53.3	33.3	53.00	36.6	11.5	53.0	9.46	9.93	Light Pink
F.	26.3	77.66	54.7	50.0	26.3	45.00	55.4	15.8	44.3	8.03	4.6	Dark Pink
H.	40.0	137.33	57.3	63.2	35.0	96.66	46.7	16.3	53.0	10.7	5.9	Pink
I.	18.0	88.66	49.3	55.4	25.3	62.00	40.9	16.6	96.6	9.76	8.9	Pink
J.	28.7	95.66	54.7	54.7	20.0	71.66	52.6	14.6	62.0	11.0	8.96	Light Pink
K.	29.7	71.66	51.7	48.3	21.6	44.66	77.2	16.6	71.0	9.4	5.23	Pink
CD 5%	14.9	27.62	NS	9.1	12.2	36.90	-	1.40	44.6	0.57	0.53	

Chapter 4

PLANT PROPAGATION AND NURSERY MANAGEMENT

Pomegranate is emerging as one of the important fruit crops for semiarid and arid areas of India with the identification of superior varieties like Ganesh, Mrudila, Arka Ruby, and Bhagwa the area under pomegranate is increasing fast. Therefore the demand for elite planting material is also increasing fast. Pomegranate is propagated vegetatively either by cutting or air layering since ages. However now a day's orchards raised from tissue culture are also very common. Nursery as a business preposition can also be as profitable as raising a crop.

Types of Nurseries According to Marketing

1. Wholesale nurseries
2. Garden centers
3. Nurseries for large-scale gardening (municipals etc.)
4. Regular nurseries

Major Factors in Nursery Planning

1. Period and dates of sale
2. Demand for specific propagation methods (seed, cutting etc.)
3. Avaibility of propagation material
4. What is the ready to sale plant (size, form etc.)

Production Stages

1. Propagation
2. Liner production
3. Growing (final product)
4. Marketing

Different Methods of Propagation of Pomegranate

Seed

Raising of Seedlings from Seeds Either *in-situ* or in Nursery

In-situ raising of seedling is necessary because it helps in development of strong root system making plant sturdy which can withstand the vagaries of monsoon better and has long life than notified plants. However, the gestation period is higher.

Another method of raising seedling is by sowing seeds in raised beds at 30x30 cm spacing and once the seedlings are of are 5-10 cm tall are transplanted to already filled polythene bags with a mixture of soil, FYM, sand in the ratio of 1:1:1,use of vermicompost in place of FYM is advocated. The soil should also de drenched with copper fungicides @ 3g / l water to prevent infection of soil born fungi. The method is easy and less expensive with high percent success with long orchard life. Mostly used for evaluation of breeding materials and germplasm material obtained from overseas. The disadvantages are long gestation period, genetic variability in the material and dormancy period in some seed is long and in some viability is very short. The seeds can be treated with bacterial culture of azotobacter @30g/kg seed or with VAM culture to improve germination and rooting.

Rooting Pattern Air Layer & Cutting

Pomegranate Air Layring

Cutting

Commercial propagation of pomegranate through cuttings is not as successful as air layering due to unsatisfactory rooting. Still in humid areas in can be successfully used. Though root augmentation through growth regulators like auxin is dependable option, its use is costly.

Semi hard wood cuttings of young healthy plants from the matured stem with light white streaks on them are more successful. 20-25 cm long cuttings are removed from the plants and immediately their leaves are removed to control loss of water from the cutting. The cuttings are then wrapped in moist sphagnum moss or newspaper if to be transported to distant places or if the nursery is far off. However the percent rooting in these cutting is found to be very low (20 %). Treating these cuttings with IBA @2000ppm for 15-20 minutes was found to be useful in increasing percent success to 60 percent. Best time for taking cutting is on the onset of monsoon. Cuttings are planted at 30x30 cm spacing on raised beds and after giving a slanting cut at the bottom of cutting one third portion is inserted into the soil to facilitate rooting and once the seedlings are of 5-10 cm tall are transplanted to already filled polythene bags with a mixture of soil, FYM and sand in the ratio of 1:1:1.Vermicompost can be used in place of FYM as it contains good amount of nutrients. The soil should also de drenched with copper fungicides @ 3g / l water to prevent infection of soil born fungi. Use of beneficial microbes is a novel approach to reduce the cost of chemical and enhance rooting /survival of the cuttings is advocated by Liberman (1993). *Agrobacterium rhizogenus* and *Trichoderma hargianum* are known to produce growth regulators, which have beneficial effect on plant growth (Riker *et al.* 1930). The unique root initiating ability of these microorganisms may set early and better rooting resulting in improvement of vigor of freshly transplanted material. Percent rooting of cuttings was maximum (76.77 %) in cv. Ganesh in *Trichoderma hargianum* along with treated cuttings, IAA (51 μg/m) Patil *et al* (2004).

Percent rooting of cutting in cv. Ganesh

Treatment	Cutting rooted (%)	Survival %	Auxin ug/ml
IBA200ppm	53.30	66.82	
Trichoderma hargianum	73.30	76.77	51
Azospirillium brasielense	23.30	74.68	32
Azospirillium lipoferum	36.20	60.07	21
Control	33.33	34.07	-
C.D. 5%	17.57	-	-

Air Layering

This is one of most successful method for large-scale propagation of pomegranate. However it is time specific i.e. can be most successful when performed after one or two showers. The success percent is as high as 80 percent even under rain fed conditions. The matured shoots on healthy plant are selected and 1.5 - 2 cm portion of shoot cambium is removed the cut portion is then covered with sphagnum moss dipped in water or in the solution prepared as below

Potassium nitrate - 0.28g / l
Calcium nitrate - 0.8 g / l
Potassium hypophosphate - 0.2g / l
Magnesium sulphate - 0.2g/ l

Ferric tartarate one ml of 0.5 % solution.

Apart from this vermiwash also can be used as it is also found to have hormonal and antifungistatic properties. Sphagnum moss can hold about 5 times of water its weight and hence can provide congenial atmosphere for rooting. The moss is then wrapped with a polythene sheet of 15x15 cm size and tied with sutli at both the ends. The air layers root in about 30 days period, when they can be separated from mother plant; their leaves are removed to prevent loss of water and easy transportation. The layers are then planted in polythene bags already filled with the soil mixture after removing the polythene sheet keeping the moss intact. Root length and weight was more in air layers compared to cuttings under semiarid rainfed conditions. Maximum root length was 27 cm in air layer as to 12.45 cm in rooted cutting. Fresh root wt. was 1.45 g as to .513 g in cutting which results in higher success in nursery after detachment of air layer from the plant as compared to the cutting.

Tissue Culture

Micro propagation is one of the improved techniques, which can be used for propagation of elite plant material on large scale under controlled conditions. In India the technique is picking up and it is becoming a industry with the opening of a trade for export of plant material. This can emerge as one of the profitable industry. In pomegranate also protocol have been standardized for tissues culture of the shrub by shoot tip culture or cotyledon callousing. Elite material was collected. Sterilization of the collected material is done with sodium hypocloride 05 % for 5 minutes and finally treating them with 0.1 % $HgCl_2$ 0.5 after successive rinsing with distilled water the sterilized material is innoculated on Murashige and Skoog media. Murkute *et al* (2002) worked on profuse callousing and differentiation of callous into whole plants in pomegranate Cv. Ganesh using leaf segment and cotyledon explants. Cotyledon explants was found to be free from lethal leaching of phenolic compound on M.S. basal medium fortified with BAP and NAA. Better performance was obtained with BAP + NAA combination cotyledon explants responded better than leaf segments profuse callusing was obtained in M.S. medium with BAP (1.0mg l^{-1}+ NAA 0.5 mg l^{-1}) and good rooting was obtained in ½ M.S. + IBA 1.0mg l^{-1} direct regeneration of shoots, roots and whole plant with intervention of callosing have also been worked at Naik *et al* (1999), use of anther wall (Morgud *et al*) and leaf segment (Omula *et al*) has resulted in production of few seedlings only. In somatic embryo genesis (Narayana and Neelambica) there was no scope for screening of variants.

Aims and Applications of Tissue Culture

1. Rapid propagation of new hybrids
2. Seed and embryo germination and rescue
3. Vegetative propagation
   i. Conventional propagation is not possible
   ii. Limited amount of stock plants
   iii. Low propagation coefficients
4. Pathogen free propagation material
5. Gene banks for crop improvement

Stages in Micro Propagation

1. Mother plant selection and preparation, (Desired, pathogen free clone).
2. Establishing an aseptic culture (Aseptic, Viable, Actively growing).
3. Production of suitable propagules (Rapid initiation and mass growth).
4. Preparation for ex-vitro growth (Hardening, Adaptation to soil).

Factors Responsible for Faster Multiplication

1. The organ (Branch, bud, meristem).
2. Ontogenesis and physiological age (Young, Adult, Juvenile adult).
3. Season during the year.
4. Size of the explant.
5. General quality of mother plant.
6. Culture conditions.
   i. Asepsis
   ii. Culture medium composition
   iii. Physical state of medium (Agar, Liquid etc.)
   iv.. Environmental conditions (Temp. Light, Atoms.)

7. Cytokinin, auxin ratio: concentration of cytokinin increased and auxines decreased for rooting concentration of rooting hormone like IBA is increased and cytokinin decreased.
8. Basic physico-chemical composition and nutritional requirement of nitrogen, amino-acids and vitamins in the media.

Economics of Pomegranate propagation

Particulars	Air-layered	Cutting		Seedling
		With IBA	Without IBA	
Polythene sheet (22.5m) 15cmx15cm pieces	Rs.360 (Rs.16/m)	-	-	-
Moss grass (5kg)	Rs.200 (Rs.40/kg)	-	-	-
Polythene bags (1000nos)	Rs.250	Rs.250	Rs.250	Rs.250
Sutli (3kg) @ Rs.25/kg	Rs.75	-	-	-
Soil mixture (1 load of trailer)	Rs.500	Rs.500	Rs.500	Rs.500
Manpower (Malies)	Rs.500	Rs.100	Rs.100	Rs.50
Laborers for maintenance (10nos)	Rs.500 (Rs.50/lab)	Rs.500	Rs.500	Rs.500
IBA @2000 ppm	--	476/ 5g	--	--
Success (%)	80%	60%	20%	50%
Total Cost	**Rs.2385**	**Rs.1826**	**Rs.1350**	**Rs.1300**
Cost/plant (Total cost/ no. Successful)	**Rs.2.98**	**Rs.3.24**	**Rs.7.50**	**Rs.2.60**

Profit from nursery (one Lakh Plants)

	Air-layering	Cutting		Seeds
		With IBA	Without IBA	
Expenditure/plant	2.98	3.24	7.50	2.60
Success (%)	80%	60%	20%	50%
Expenditure (one Lakh plant)	2,98,000	3,24,000	7,50,000	2,60,000
No. of successful plants	80,000	60,000	20,000	50,000
Gross income @Rs. 10/plant	8,00000	6,00000	2,00000	2,50000
Net profit	**5,02000**	**2,76000**	**Nil**	**Nil**

Plant grown from seed sold @ Rs. 5 per plant.

Disadvantages of the Method

1. Costly method in Banana single plant is sold @ Rs. 10/- whereas, normal rhizome is available @ Rs. 0.5/- to Rs. 1/- per rhizome.
2. Sometimes it is difficult to get success e.g. Date palm
3. Failure of micro propagated plants is higher under field conditions due to harsh climatic conditions.

Green House Propagation

High quality reliable green hoses with controlled atmospheric conditions are required in achieving desire success. Temperature control is most commonly manipulated. Mechanical ventilation along with use of fog cooling and adequate shading systems are provided. In pomegranate Mist propagation increases percent success in cuttings to 80 percent but then the cost of establishing mist chamber is around Rs.1.5 Lakh for a standard green house.

Chapter 5

CULTURAL MANAGEMENT

Soil and Climate

Commercial pomegranate orchards are confined mainly to Mediterranean, and semi-arid regions of the world. It is hardy fruit, which can be grown successfully, even in low fertile soils. Although pomegranate of high quality can be grown only where there is a cool winter and a hot dry summer, the trees grow under wide range of climatic conditions. It can be grown from the plains to an elevation of about 1829 m. The tree can withstand frost but gets injured by temperatures below 11.1°C.

In areas of low winter temperatures the tree is deciduous but under tropical and sub-tropical conditions, it is evergreen or partially deciduous. The tree cannot produce sweet fruits, unless the temperatures are high for sufficiently long period. The quality of fruit is adversely affected in humid climate. It is hardy plant and can withstand considerable amount of drought but does better when irrigated regularly.

The pomegranate can be grown on diverse soil types. The tree can be grown on soils, which are considered unsuitable for most other fruit trees. It can be grown quite well at the foothills or at high elevations. It can tolerate soils, which are and slightly alkaline. The deep loamy or alluvial soils are ideal for its cultivation though it can be grown in medium or light black soils of minimum 60 cm depth.

Preparation of Land and Planting

The land selected for cultivation of pomegranate is ploughed deeply with a furrow turning plough followed by 3-4 fine tillage, the layout is done according to square or hexagonal system. Planting is done in previously dug pits of $1m^3$ size. The pits should be filled with 2-3 baskets of FYM, 50g diammonium phosphate (DAP) same quantity of sand mixed thoroughly before the on set of the rainy season. Planting should be done during monsoon for better and rapid establishment of the plants. The planting is done 3-6 m apart. In Maharashtra, spacing of 3 to 5 m is generally followed. In south India a spacing of 2 m is common. However, the trees should be given a spacing of 5-6 m in poor, shallow soils of the Deccan and 6-7 m in the rich alluvial soils of Gujarat and North India.

Spacing

In Pomegranate earlier recommendation for spacing was 5x5m but with the mechanization of cultivation and adaptation of drip irrigation the spacing adopted by the farmers are 14x14 feet, 12x12 feet, 15x10 feet 10x10 feet, 12x10 feet.

Bahar/Resting Treatment to Regulate Fruiting

Bahar treatments like root pruning, root exposure, withholding water, defoliation of plants by hand or by chemical defoliation etc., are practiced to induce moisture stress, so that the plants drop their leaves and the growth will be controlled. The main object of these treatments is to regulate the crop by forcing tree to take rest and produce profuse flowering and fruiting during any one of the three bahars. The treatment also helps in getting uniform and good quality fruits with maximum production and less incidence of pests and diseases. Normally, 2 bahar are practiced in Maharashtra and Dholka area of Gujarat. Once the trees shed their leaves, recommended dozes of fertilizers are applied depending on the selection of the bahar, however, basal dose of F.Y.M, phosphorus and potash with ½ dose of N are applied on the onset of monsoon.

Allowing a particular bahar or flowering depends on market rate of fruits and arrival of produce to the market, water availability to the plants, type of soil, insect pest menace and weather conditions during crop growth period. Under tropical south Indian climatic conditions, pomegranate flowers continuously throughout the year. In central and Western India, There are three distinct seasons of flowering i.e., Ambe bahar (January-February), Mrig bahar (June-July) and Hast bahar (October). Ambe bahar is most commonly practiced by the growers of Maharashtra because of high yield, as compared to other bahars. However assured irrigation is a must to take this bahar. In Karnataka, flowering is observed in June-October and March months. Under north Indian conditions, the trees remain dormant during winter and flower only during spring. During monsoon again, a flush of flowers are observed from April to June. Under Delhi conditions, depending on cultivars grown, the flowering may be once in a year or twice. In Himachal Pradesh, where there is temperate climate, the flowering is seen during middle of April.

For Mrig bahar, plant growth has to be suppressed during December to April by withholding water. This is generally followed in South India. By doing so, the trees shed their leaves in March and remain dormant till May.

Flowering Plant

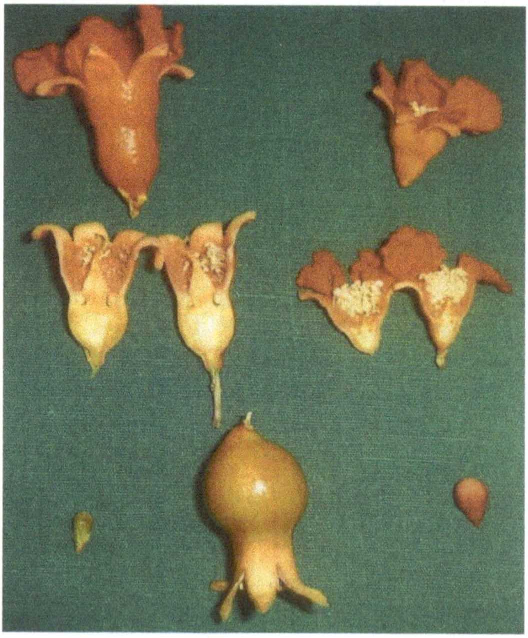

Hermaphrodite Flower and Flower with Rudimentary Gynosium (Different Stages of Development)

Then the land is ploughed followed by application of manures and fertilizers. Plants are irrigated till the rainy season and the fruits are harvested during October-November. In North-West India, irrigation is withheld from December to May, then regular digging is taken up followed by application of manures and fertilizers during June along with normal irrigation till the rains start. Plants starts flowering by June and fruits are harvested during October-December. To get good size fruits, flower and fruit thinning is practiced. To induce flower and fruit drop, application of 2000 ppm Ethepon and 500 to 3000 ppm Alar have been tried. Fruitlet thinning has also helped in getting good size fruits. At Rahuri, fruit size in Muscat pomegranate during Ambe bahar progressively increased when the number of fruits per tree was decreased from 70 to 20 and the gross income was highest when only 50 fruits were retained/tree.

Effect of bahar on Flowering in pomegranate cv. Ganesh.

	Mrig				Hasta			
	1986	1987	1988	Mean	1986	1987	1988	Mean
No of Herma -	219.5	236.57	301.7	252.59	249.6	304.8	284.5	279.63
phrodite flower	(28.88)	(12.29)	(28.32)	20.21	(22.13)	(34.10)	(28.44)	27.76
No of Staminate	468.57	1054.71	568.6	697.29	788.8	420.4	531.2	580.13
flowers	(61.63)	(54 82)	(53.37)	55.80	(69.95)	(47.03)	(53.11)	57.59
No of unopened	72	617.14	195	294.71	89.2	168.6	184.5	147.43
flowers	(9.47)	(32.08)	(18.30)	23.58	(7.91)	(18.86)	(18.44)	14.64
No of flowers	540.57	1671.85	763.6	992.01	878.0	489.0	715.7	694.23
dropped	(71.12)	(86.91)	(69.14)	79.38	(77.86)	(54.71)	(71.55)	68.92
Total flowers	760.07	1923.57	1065.3	1249.6	1127.6	893.80	1000.20	1007.2

Annon (1996)
Fig. in parentheses are percent values

Under semiarid raifed condition of Central Gujarat there was lot of variation in flowering during Mrig and Hasta bahar. Though highest total no of flowers were produced in Mrig bahar, (1249.6). Total flower drop was highest (28%), maximum hermaphrodite flower production was observed in Hasta bahar (27.76%) and reduced flower drop (68.92%) compared to Mrig bahar resulted in higher fruit set and ultimately yield. (Raturi and Hiwale 1991).

Fruit Development

In Mrig bahar fruit development in cv. Ganesh under semiarid rainfed conditions of central enjaret indicated that after fruit set on the onset of monsoon it took around 120 days for the fruit to reach the maturity.

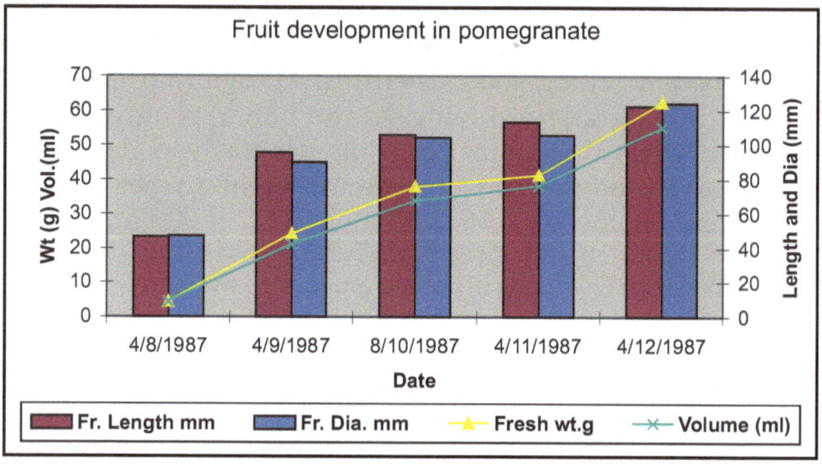

Fruit length and diameter increased at a faster rate initially up to 60 days and then increased slowly. Fruit weight and volume increase constantly till maturity as below figure. Similar trend was observed in hasta bahar crop also.

Chapter 6

WATER MANAGEMENT

Irrigation/ Drip/ Fertigation

Pomegranate is considered to be drought hardy fruit crop. Supplemental irrigation, however, is necessary for realizing higher fruit yield and better quality. Irrigation is also used to regulate cropping in pomegranate and the total water applied therefore depends on the desired bahar. In Gujarat and Maharashtra, where Ambe bahar is taken, regular irrigation is required during March to July and irrigation is stopped after the fruit harvest until January. For Mrig bahar crop irrigation is not required from July to September. However, in case of prolonged break in monsoon one or two supplementary irrigation at an interval of 10-15 days during October to December are required. In hasta bahar (September flowering) prolonged rainy season helps in reducing requirement for irrigation 3-4 irrigation at monthly interval after cessation of rainfall are must to obtain good fruit size and higher yield.

Comparison of Traditional and Drip Irrigation System

- Water is flooded with no control
- Water is applied without taking into consideration the crop water requirement
- Whole field is wetted instead of to the plant
- Water is applied at a faster rate with in short time
- Less area is irrigated with large volume of water
- Water use efficiency is just 30 to 35 percent
- Weed population is more
- Limited production
- In undulating land it is difficult to irrigate the plants by this method
- It is not possible to apply fertilizers along with water in this method
- It is not possible to reduce fertilizer dose to be applied
- There is slow degradation in soil, soil becomes saline if irrigation water is saline.

Drip Irrigation

- Controlled application of water
- Water can be applied according to requirement
- Water is applied with in the periphery of the plant
- Water is applied at low pressure and for long time
- More area is wetted with less water
- Irrigation efficiency is 90 to 95 percent
- Less weed infestation
- Definite increase in productivity of the plant
- It is possible to irrigate even in undulating land
- Less expenditure on management
- Fertigation is possible
- Reduction in 25 to 30 percent fertilizer requirement
- Maintenance of sustainability of soil is possible
- It is possible to grow crop even in light soils with low fertility.

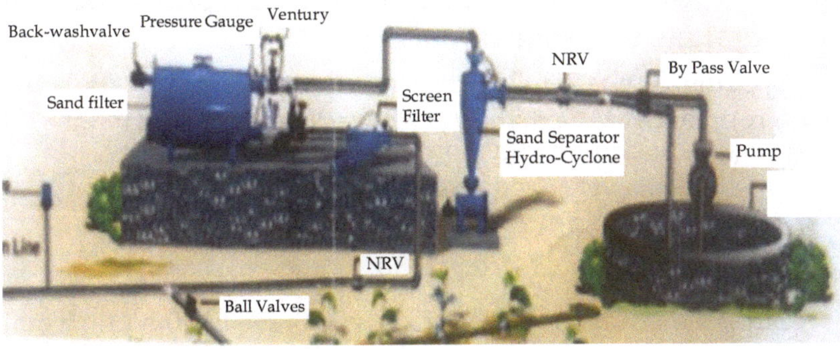

Drip Irrigation System

Crop Coefficient for Different crops

Name/ Age of tree	1st yr	2nd yr	3rd yr	4th yr	5th yr
Pomegranate	0.4	0.4	0.6 - 0.65	0.6 - 0.65	0.60 - 0.65
Ber	0.4	0.4	0.6 - 0.65	0.6 - 0.65	0.60 - 0.65
Fig	0.4	0.4	0.6 - 0.65	0.6 - 0.65	0.60 - 0.65
Aonla	0.4	0.4	0.5	0.6	0.65 - 0.70
K.Lime	0.4	0.4	0.5	0.6	0.70 - 0.75
Guava	0.4	0.4	0.5	0.6	0.70 - 0.75

Formula for Calculation of Quantity of Water to be Applied per plant/ day

A = Evapotranspiration /Day x Pan Coefficient (0.7to 0.8)
B = Crop Coefficient (From table)
C = Crop Canopy (At 12 noon)
D = Spacing
E = Area wetted= Crop canopy/ Spacing

Growth Parameters of pomegranate Ganesh

Year	Pl. Height m	Stem Dia mm	Spread NS m	Spread EW m	No. of Fruit Set/ plant
3rd	1.4	14.1	1.35	1.4	45
4th	1.76	17.9	1.58	1.7	75
5th	2.36	36.4	2.35	2.2	120
6th	3.03	48.4	2.8	2.7	80

Quantity of Water to be Applied per plant/ day in Pomegranate Cv. Ganesh

Spacing 5x 5 =25m2
Crop canopy/ Spacing (yearly)= Wetted Area/Plant
3rd year = 1.35x1.4/ 25=0.075
4th = 1.58x1.7/25=0.11
5th = 2.35x2.2/25= 0.21
6th = 2.8x2.7/25=0.30

Suppose A= 8.5mm, B=0.75, C=0.70, D=13.40 sq.m. Total water requirement /day will be 59.80 liters.

Daily Water requirement liters/Plant in semi arid rainfed areas of Central Gujarat

	3rd year	4th year	5th year	6th year
January	3.64	6.17	10.2	15.79
February	4.43	6.5	12.41	19.21
March	6.09	8.94	17.07	26.42
April	8.85	12.98	24.79	38.36
May	10.32	15.15	28.92	44.75
June	8.31	12.19	23.25	36.02
July	5.11	7.49	14.3	22.13
August	4.57	6.7	12.78	19.79
September	4.61	6.76	12.92	19.98
October	3.94	5.79	11.05	17.11
November	3.76	5.52	10.55	16.33
December	3.31	4.86	9.29	14.38

In case of flood irrigation during winter, irrigation should be given at the interval of 12-15 days and during summer at the interval of 5-7 days depending on the bahar. However with the limitation of water availability now a days drip irrigation / fertigation is becoming popular. According to one of the estimates there is 98 percent increase in production with a saving of water to the tune of 45 percent. Total area brought under drip in Maharashtra is around 27,217.39 ha (10.26%), which is the highest in India.

Mulching and Drip Irrigation in Pomegranate

Jagannath *et al.* (1999) in their studies on drip irrigation and mulching maximized water use efficiency of pomegranate cv. Jyothi. Total flowers and hermaphrodite flowers were significantly higher at 0.70 E.pan and 200 gauge polyethylene mulch. The ratio of hermaphrodite to staminate flowers was maximum (2.54) in the interaction, three emitters/plant at 0.70 E.pan irrigation with 400 guage polyethylene. Fruit set (56.39 and 51.94) and fruit per plant (55.36 and 45.89) were significantly higher with 0.70 E.pan replenishment along with 400 guage black polyethylene mulch, respectively.

Fertigation in Pomegranate

Application of fertilizers through drip is gaining importance because of following reasons

- Reduction in the cost of application of fertilizers along with water.
- Balanced application of fertilizers based on the hunger of the plant.
- The fertilizers are applied near the active root zone of the plant which helps in absorption by the roots
- High fertilizer use efficiency can be achieved.
- Saving in the total quantity of fertilizer applied.
- Fertilizers are applied based on the stage of the crop.

Application of Fertilizers

New Growth, Flowering : In the beginning for achieving uniform growth 20:20:20 grade should be used. For uniform and profuse flowering phosphorus should be applied 3 to 4 times more than nitrogen quantity of potash is lowered. There fore after shoot growth till fruit set the ratio of fertilizer applied should be either 12:61:0 or 13:40:13.

Fruit Set and Development

To keep higher number of leaves on the plant so that maximum photosynthetic activity takes place during fruit development the ratio of NPK should be 2:1:3. For example 16:8:24 or 13:5:26.There is increased demand for calcium at this stage, there fore calcium nitrate should be applied through drip. It will help in reducing the fruit cracking and improve fruit size.

Application of liquid fertilizers

Liquid fertilizer	Total fertilizer required (kg /acre)	Weekly dose applied (Kg/ week/ acre)	Time of application
20:20:20 +urea	10 +10	10 +10	First irrigation
20:20:20 +12:61:0+ urea	15 +50 +22	3 +10 + 4.40	6 to 25 days (5 weeks)
13:40: 13+ 16:8:24	50 +25	10 +5 (5weeks)	36 to70 days (weeks)
13.00:45	50	12.50 (4 weeks)	71 to 98 days (4weeks)
00:00:50	50	10	91 to133 days (5 weeks)
06:12:36	25	12.50	After 134 days

Bachary and Bhosale (2005).

Yield and yield-contributing parameters of pomegranate under fertigation

Treatment	Fruit No. /tree	Fruit Wt. (g)	Fruit size (cm)	100 aril wt. (g)	Yield (t /ha)	Water use efficiency (Kg/ha/mm)
Conventional fertilizers						
T1 (100%RDCF + SL)	73.45	191.00	7.00	31.27	8.71	7.71
T2 (100%RDCF + DI)	76.50	200.00	7.46	33.60	9.54	12.08
T3 (100%RDCF, N drip)	80.25	209.00	7.72	36.77	10.36	13.12
Solid soluble fertilizers						
T4 (100% NPK + DI)	84.80	226.00	8.22	42.95	11.88	15.05
T5 (70%N, 80%P&K + DI)	82.60	220.00	8.10	40.20	11.27	14.28
T6 (70%NPK + DI)	81.50	212.00	7.85	37.87	10.81	13.70
T7 (50%N, 80%P&K + DI)	75.15	145.00	7.23	33.40	9.14	11.58
T8 (50%N, 70%P&K + DI)	75.25	191.00	7.81	32.75	8.90	11.27
C. D. at 5%	6.21	0.02	0.57	4.20	1.12	-

* (Firke and Kumbhar 2002)

At M.P.K.V Rahuri, combined application of fertilizers by conventional method and through drip revealed that an yield of 11.88 t./ha obtained due to application of 100 percent NPK (recommended dose) as solid soluble fertilizers through drip which was significantly superior over 100 percent Recommended dose of conventional fertilizer and was at per with 70% N, 80% P and K and 70% NPK. Thus the maximum fertilizer saving (N, P and K) was 30 percent as compared to conventional fertilizer. The yield contributing parameters were significantly influenced by different levels of solid soluble fertilizer over conventional application. Fertigation levels did not significantly influence the juice quality and organoleptic evaluation for colour and taste of pomegranate (Firke and Kumbhar 2002).

Effect of drip irrigation on yield and quality of fruits

Treatment (liters/hr)	Juice (%)	Cracking (%)	Fruit weight (g)	TSS Brix	Yield (kg/plant)
4	46.3	25.8	204.9	18.0	18.8
8	53.6	18.9	319.4	18.4	28.2
12	53.0	18.6	314.7	17.4	30.1
Control	44.4	30.1	160.2	19.3	17.7

Prasad and Bankar (2003) reported that with drip irrigation the fruits become ready for harvesting 4-5 months after fruit setting. They are harvested when their rind color turns yellow and pink to red. Fruits give a 'crunch' sound when pressed. Thus, an yield of 60-100-fruits/tree maybe obtained. First week of December to February end is right time for plucking of fruits. On an average, 25-30 kg fruits/plant may be harvested under arid condition.

Fruit Ripening

At this stage requirement of potash is higher than that of N and P there fore 0: 0: 50 +18 % sulphur is applied. In light and soils deficient in nitrogen are however fertilized with 13: 0: 45 grade fertilizer. After initial harvest of big size fruits a second dose of fertilizer is applied to increase fruit development in underdeveloped fruits through 6: 12: 36 grade fertilizer.

Chapter 7

NUTRIENT MANAGEMENT

Manures and Fertilizers

Even though pomegranate grows well in soils of low fertility, production can be increased by application of manures and fertilizers. Initial soil analysis is desirable for proper scheduling of fertilizers. The nutritional recommendation depends on fertility of the soil and also the age of the plant. For young plants (2-3 months old) application of 200 g Neem cake along with 3-4 kg FYM/plant is recommended. After 3 months, each plant may be given 250 g DAP along with 250 g Neem cake and 5 kg FYM. Again after 9 months, application of 500 g DAP, 250 g Potassium sulphate, 1 kg Neem cake and 10 kg of FYM/plant is recommended.

The first harvest is desirable when the plant is around 12-16 months old. There are three distinct periods of fruiting in pomegranate called "bahars". Before each bahar, the plant needs one to one and half months rest. During this time, the nitrogen level in the soil should be low. Once the fruits are harvested the plants have to be supplied with first dose of fertilizers along with a fungicide and insecticide spray. The recommended dose of fertilizer per plant is 20 kg of FYM, 1 kg (7:10:5), 1 kg DAP, 1 kg Neem cake, 100 g magnesium sulphate and 10 kg of borax. After rest period, apply 250 g ammonium sulphate along with 250 g of 19:19:19, 500 g 7:10:5, 500 g Neem cake and 500 g Murate of potash per plant. When the plants are 4-5 year old, the fertilizer dose has to be increased to one and half times. In general, basal dose of about 25-35 cartloads of FYM per hectare besides the recommended dose of NPK must be applied during January, June and September months in equal proportions. From 4-5 year onwards nitrogen should be applied in 2 split doses, first during irrigation after bahar treatment, the second after 3 weeks. Full dose of P and K should be applied as a single dose. The requirement of chemical fertilizers also varies from place to place. In Gujarat, 375 g N, 375 g P and 375 g K per tree per year has been recommended. Under Udaipur conditions, a four year old tree needs 720 g N, 200 g P and 220 g K per year. Under Yercaud conditions, young plants are started with 50 g N, 200 g P, 100 g K and as the plant becomes old, the dose should be increased to 125 g N, 1000 g P, 500 g K. Under Jodhpur conditions 700 g N per plant is ideal for 10-year-old plants of 'Jalor Seedless' variety.

Under Rahuri conditions 1 ½ to 2 year old plants need 250 g N, 125 g P, 125 g K, 2 ½ to 3 year old plants need 500 g N, 125 g P, 125 g K, 4 ½ to 5 year old plants need 500 g N, 250 g P, 250 g K and 6-7 year old plants need 625 g N, 250 g P, 250 g K (Anon, 1979).

At IIHR, Bangalore, application of 500 g N + 250 g P + 125 g K/plant/year to Bassein seedless variety of pomegranate gave the highest yield of 8.2 kg fruits from 3 year old plant (first crop) and application of 500 g N, 250 g P and 250 g K/plant/year to Cv. Ganesh gave the highest yield of 78.67 kg (50 fruits/plant) during fourth year of cropping. (Prassana Kumar and Dhandhar 1996).

Critical levels of Nutrients in Pomegranate leaf composition

S. no.	Nutrient	Concentration (%)	S. no.	Nutrient	Concentration (%)
1	Nitrogen	2.5	6	Sulphur	0.18
2	Phosphorus	0.2	7	Iron (ppm)	95
3	Potash	1.47	8	Manganese (ppm)	45
4	Calcium	1.30	9	Zinc (ppm)	22
5	Magnesium	0.70	10	Boron (ppm)	30

Pomegranate orchard survey in Maharashtra

Place	PH	E.C. dsm[1]	Org. Carbon	K ppm	Zn ppm	Mn ppm	Cu ppm	Fe ppm	B ppm
Pune	8.05	0.81	0.53	174	0.51	8.83	0.25	6.42	0.47
Solapur	8.35	1.42	0.24	189	0.46	8.13	0.36	5.3	0.48

* Edward Raja, 2008.

Leaf composition (Mean of Six orchards)

Place	Zn (Ppm) Young leaf	Mn (Ppm) Young leaf	Fe (Ppm) Young leaf	K (%) Old leaf	Mg (ppm)	B (Ppm) Young leaf
Pune	12.33	73.5	161.0	0.79	0.32	22.0
Solapur	12.00	44.66	99.33	0.59	0.31	27.0

* Edward Raja, 2008.

Based on the survey it is recommended that combined foliar application of 0.3 percent Zinc sulphate and 0.1 percent boron at prebloom, bloom and post bloom stages should be done to augment the deficiency of these nutrients. The soil application can also be done by incorporating boron (10ppm) and Zinc (50ppm) in compost 3 months before its application. Though the soils are reach in calcium the cracking may be due to combined deficiency of calcium and Boron. Hence two sprays of calcium chloride @0.5 percent and Boric acid @0.1 percent is recommended at 15 days interval to avoid cracking.

Commercial micronutrients formulations are in use to increase fruit growth, yield and quality. Two sprays of 0.33 percent Macroliq proved to be the best followed by 1 percent Macroliq. Soil application of 12.5 g boron and 45 g zinc sulphate per tree per year gave fruits with higher sugar content. Application of 3 sprays of ferrous sulphate (0.4 %) + manganese sulphate (0.3 %) + boric acid (0.2 %) along with Zinc sulphate (0.3 %) at monthly interval given before flowering, at full bloom and fruit set stage, increase tree growth, yield and fruit quality in Cv. Ganesh.

Organic Fertilizers

The investigation carried out on Pomegranate cv. Ganesh on suitability of bahar under rainfed semiarid conditions revealed that *Hasta* bahar was the best Raturi & Hiwale, 1991. Importance of organic farming in India is increasing due to rising threat to natural resources and food security from contamination. With the advent of green revolution indiscriminate use of chemical fertilizers, pesticides, weedicides had adversely affected the soil fertility, productivity and quality of produce. Soil and water pollution with the heavy metals released by fertilizers and pesticides resulted in health hazards. The organic farming system relies on crop rotation, crop residues, animal manures, and legumes in intercropping system, green manures organic cakes as byproducts of plants and biofertilizers there by increasing the sustainability of soil and plant.

Benefits of Organic Farming

- Sustainability of soil health
- Soil moisture conservation
- Reduced environmental hazards and improved public health.
- Increased farm income through sale of organic produce at higher rate
- Improvement in quality of produce.

With the objective of reducing the use of chemical fertilizers and increasing the use of organic fertilizers, without compromising the production potential of pomegranate a trial was conducted at CHES, Vejalpur. In the study, six-treatment combination of organic and in organic fertilizer along with the recommended doses for semi arid region were applied on the onset of monsoon in 4 and 5 year old plant in cultivar Ganesh. Leaf sampling was done by collecting 8th pair of leaf from growing tip as suggested by Bhargava and Dhandhar (1987).

Fruit retention and yield per plant were significantly influenced by various treatments. Maximum fruit set (96.50 fruits / plant), fruit retention (57 fruits / plant) and yield (10.75 kg / plant) were recorded in treatment, application of Nitrogen fifty per cent through FYM, twenty five percent through castor cake and urea each. Similar results were reported by Shinde (1977). Phadnis (1974) also recommended mixed doses of organic and inorganic fertilizers to obtain maximum fruiting and yield. Wavhal *et al* (1985) also recommended application of nitrogen along with *azotobactor* for obtaining higher production in pomegranate cv. Ganesh.

Fruit weight and fruit length were significantly influenced by all the treatments except fruit diameter and T.S.S. Fruit weight (188.75g), fruit length (69.72mm), fruit diameter (66.15mm) and T.S.S. (16.90 ^{0}Brix) were highest in application of nitrogen fifty per cent through FYM and twenty-five percent through castor cake and urea. Bankar *et al* (1990) reported that increase in doses of N did not affect the fruit size and T.S.S. in pomegranate.

Nutrient analysis of leaf samples collected one month after fertilizer application showed significant influence in respect of Nitrogen and Potassium, and non-significant differences in respect of Phosphorus content. Maximum leaf N content (3.15%) was observed in full Nitrogen application through Urea. Maximum Phosphorus (0.14%) and Potassium (0.74ppm) content were observed in treatment application of N through fifty per cent F.Y.M, and twenty five per cent through castor cake and urea each. Similar results are reported by Wavhal (1981). Oil cakes though insoluble in water are quick acting organic manures their nitrogen quickly becomes available to the plants in about a week or ten days (Guar *et al.*1971).

Effect of organic and inorganic Source of N application on Fruit and Nutrient composition of pomegranate cv. Ganesh

Treatment	Fruit wt. g	Fruit length mm	Fruit dia. mm	TSS °Brix	N (%)	P (%)	K (ppm)	Fruits retained /plant	Yield Kg /plant
100%FYM	179.00	65.70	62.57	16.20	2.74	0.13	0.69	46	8.23
100%Castorcake	178.50	63.62	63.50	16.30	2.93	0.13	0.68	46	8.20
50%FYM + 50%Castorcake	185.25	65.72	63.27	16.35	2.82	0.12	0.73	48	8.84
50%FYM + 50%Castorcake	188.75	69.72	66.15	16.90	2.95	0.14	0.74	57	10.75
50%FYM + 50%Urea	175.50	64.15	62.95	16.35	3.02	0.11	0.74	46	8.11
100% Urea	169.75	60.45	60.97	16.05	3.15	0.11	0.66	33	5.80
CD at 5%	8.21	3.81	NS	NS	0.21	NS	0.07	12.05	2.19

Hiwale *et al.*, 2006
8th leaf pair from tip used as indicator tissue for all analysis.

Biofertilizers

Biofertilizers are preparations containing live or latent cells of efficient strains of nitrogen fixing, phosphate solublizing or celluolytic microorganism used for application to seed, soil or compost with the objective of increasing the no. of such organism which will accelerate the process of nutrient availability to the plant. They are very efficient even when applied in small quantity. Their major advantages are

- Supplementation of nutritional requirement of the plant
- Addition of 20-200 kg N per hectare as well as can mobilize 30-50 kg P_2O_5 per hectare.

- Liberation of growth promoting substances.
- Suppress the incidence of pathogens.
- Improvement of soil health.
- Increase in the crop yield by around 10-50 percent.

Careless management of the orchards and ruthless use of chemicals have already showed their deleterious effects on crop performance in Pomegranate in Andhra pradesh, Karnataka and Maharashtra. Due to excessive fertilization and water application, the orchards are succumbing to the infection of pest and diseases, soil health is also spoiled. Unless and until balanced management is practiced; it will not be possible to save the orchards from declining. Organic fertilizers play a significant role in maintaining the sustainability of the soil, improving soil physico-chemical properties, micro fauna activity and hence an experiment was laid out with different bio-fertilizers to find out their effect on sustainability and productivity of pomegranate cv. Ganesh.

Biofertilizers had no influence on vegetative parameters of pomegranate. Azospirillium @ 100g per plant resulted in higher N content of leaf as well as higher fruit set and retention, however, fruit weight and fruit size and yield was maximum in combined application of PSB+ Azospirillium @ 50g per plant. Use of biofertilizers like Azotobacter has reduced the requirement of fertilizers. Application of 250 g Azotobacter culture with 100 g N has given the same effect as application of 300 g N per plant, i.e., saves 200 g N per plant. Nutrient analysis of leaf samples collected one month after fertilizer application showed significant influence in respect of Phosphorus and Potassium where as, no significant differences in respect of Nitrogen content in various treatments.

Maximum fruit set (144.75 fruits /plant) and fruit retention (83 fruits / plant) were recorded in Azospirillium 100g/ plant and yield (11.08 kg / plant) in 50g Azospirilium + Phosphate solubilizing bacteria culture owing to maximum fruit weight

(213 g fruit). Phadnis (1974) also recommended mixed doses of organic and inorganic fertilizers to obtain maximum fruiting and yield. Wavhal *et al* (1985) also recommended application of nitrogen along with azotobacter for obtaining higher production in pomegranate cv. Ganesh. Shinde (1977) reported similar results.

Effect of bio-fertilizers on yield and yield attributes of Pomegranate cv. Ganesh

Treatment	No. Fruit set/pl.	No. Fruit retained/pl.	Fruit Yield kg/pl.	N (%)	P (%)	K (%)
PSB Culture	123.50	73.50	8.37	2.88	0.156	0.68
Azospirillium	144.75	83.00	8.29	2.67	0.177	0.71
PSB+ Azospirillium	135.50	78.25	11.08	2.92	0.178	0.84
VAM Culture	136.50	79.80	8.56	2.79	0.173	0.76
Control	72.50	51.00	7.16	2.52	0.121	0.70
C.D. 5 %	34.77	14.18	0.872	NS	0.017	0.06

Hiwale *et al.*, 2006

Effect of bio-fertilizer application on quality of Pomegranate cv. Ganesh

Treatment	Fruit Wt. (g)	Dia. (mm)	Length (mm)	Aril Wt. (g)	Skin Wt. (g)	Juice (%)	TSS
PSB Culture	201.25	67.62	76.41	136.75	62.50	72.57	15.17
Azospirillium	173.75	67.27	74.75	106.25	53.25	73.89	14.65
PSB + Azospirillium	213.00	70.12	78.87	137.37	66.62	74.62	16.05
VAM Culture	188.75	98.02	72.75	117.00	58.12	73.29	15.86
Control	153.25	67.50	68.75	95.12	43.87	74.46	15.43
C.D. (5%)	24.97	NS	NS	24.30	6.59	NS	0.86

Hiwale *et al.*, 2006

Fruit weight, aril wt., Skin wt. and TSS were significantly influenced by all the biofertilizers. Fruit weight (213g), fruit length (78.87mm), fruit diameter (70.12mm) and T.S.S. (16.05 ^{0}Brix) were highest in treatment application of 50g Azospirilium + Phosphate solubilizing bacteria culture, which may be due to beneficial effect of P solubilizers with Nitrogen Fixers. Bankar *et al* (1990) reported that increase in doses of N did not affect the fruit size and T.S.S. in pomegranate. Ram and Rajput (2000) observed good response of Azotobacter in Guava under U.P. condition. Day *et al.* (2005) reported beneficial effect of P solubilizers with Nitrogen fixers in guava L-49.

Leaf sampling was done by collecting 8th pair of leaf from growing tip as suggested by Bhargava and Dhandhar (1987). 20 g of plant material was ground in mixture grinder and 1 g sample was taken for N, P, and K analysis from the oven dried leaf samples. Leaf nutrients were analyzed as per the procedure suggested by Yoshida *et al.* (1972).

Application of biofertilizer influenced the Nitrogen content and maximum N content (2.92%) was recorded in combined application of Phosphate solubility bacteria with Azospirillium. Similarly maximum P (0.178%) and K content (0.84%) was recorded in the same treatment.

Chapter 8

HIGH DENSITY ORCHARDING AND CANOPY MANAGEMENT

High Density Orcharding

India is the second largest producer of the fruits (41.5 million tones) in the world. Still the productivity of almost all the fruits in India is very low (12 tones /ha) as compared to other fruit growing countries of the world. It is a major cause of concern to our scientists, extension personnel and orchardists. Many reasons may be attributed to the low productivity and to solve this problem our scientists have to develop high yielding varieties/hybrids, resistant to the biotic and abiotic stresses and develop better agro-techniques for realizing the maximum potential of the fruit cultivation in India. Though, we have achieved this distinction, but the per capita consumption of the fruits in our country is still 80 g/day due to our huge population, which is far below from the recommendation of 120g/capita/day. There is little scope to convert the existing area to fruit cultivation owing to the fact that our first requirement is to have sufficient food. There fore the major area expansion will be on marginal degraded land where arid and semiarid fruits play a major role, as they can be grown on these lands successfully as they are drought hardy can with stand biotic and abiotic stresses and salinity better. Therefore there is need to increase plant population/unit area i.e. giving high density orcharding.

Why High Density Planting

1. The available land area for the fruit cultivation is becoming a limiting factor.
2. Rapid urbanization, Fragmentation of land holding, Industrialization.
3. High management cost.
4. Restriction on water use.
5. Labour problems.
6. Maximum possible returns with the minimum costs in the shortest possible time.

Basic Approaches in the High Density Planting

1. Accommodation of the maximum possible number of the plants per unit area to get the maximum profit per unit of the tree volume without impairing the soil fertility status.
2. Availability of a dwarf plant is the first prerequisite for establishing any high-density orchard. It may be achieved by different approaches like use of the dwarf cultivar, dwarfing rootstock/interstock and with the use of the growth retardants.
3. Closer spacing.
4. Growth regulation by training and pruning.
5. Use of plant growth retardants.

6. Use of the mechanical devices.
7. Cropping system models and crop management practices.

Under the circumstances, the concept of the high-density orcharding has become extremely significant. This concept has been successfully applied in a number of the arid and semiarid fruits by using various techniques in India. Accordingly, efforts have been made in the different parts of the world to find out the optimum spacing for the different fruit plants along with the related technologies so as to derive the maximum return per unit area.

High-density orcharding results in early bearing and also helping in minimizing weed problem. However, the productive life of orchard is reduced due to dense canopies, which will not allow sunlight penetration resulting in slow decline. This problem can be overcome by removing plant in alternate row. The other methods, which can be used in overcoming this problem, are training and selective pruning. Productivity of almost all the fruits in India is low as compared to other fruit growing countries of the world. To over come this problem in pomegranate high-density orcharding in cv. Ganesh has been standardized under Maharashtra condition. To obtain higher production per unit area maximum number of plants were accommodated per hectare. High destiny plantation at 5*2.2 m under semi arid condition resulted in 2.5 times higher yield as compared to normal spacing of 5*5 m in 6 to 7 year old orchard.

Spacing (m)	Yield q/ha.	
	1981	1982
5*2	66.65	85.69
5*3	49.76	61.96
5*4	36.30	36.35
5*5	27.46	36.64
C.D. 5%	11.15	25.27

Anon (1996).

Canopy Management

Manipulation of tree vigor and maximum utilization of sunlight, results in increased productivity and quality. Basic principles of canopy management are maximum utilization of light. Avoidance of built up of microclimate congenial for the diseases and pests. Helps in convenience in carrying out cultivation operations. Also the inter space can be utilized for raising annuals as intercrops.

Training

Pomegranate plant is bushy in nature and throws considerable number of shoots near the ground level. Retaining all these shoots at the base increases crowding of tree frame and incidence of shoot borers and diseases. Considering these factors, keeping 3-4 stems from the base was found satisfactory for maintaining good productivity as well as proper health of the plant.

Training is done only to give shape to the tree. At the time of planting all the side shoots are removed. Young plants are allowed to grow into a bush with number of main shoots arising at the ground level. Then, they are properly trained to form a single stem with a number of well-distributed scaffold limbs. The plant is topped at a height of 60-70 cm and the side shoots 4 to 6 in number well distributed along the main stem are allowed to arise from about 40-50 cm from the ground level. As and when the plant becomes too thick, thinning out and removal of very low branches are followed. The plants are supported by trellises to provide support to the branches thereby, avoiding branch breaking. Training is done systematically during initial 3-4 years of plant growth. This helps to get uniform plants in the orchard. Once plants attain 3-4 years of age, the plant growth is tremendous. Therefore, to control the growth, to avoid cris crossing of twigs and also to allow the air and light to the canopy, training is necessarily taken up. Sometimes non-bearing twigs and water shoots along with dried and diseased twigs have to be removed by pruning.

New Plant

1st Year Old

2nd Year Old

3rd Year Old

4th Year Old

5th Year Old

Training in Pomegranate

(If pruning is not effected manually, it can be achieved by using 1000-ppm spray of Lihosin once or TIBA 500ppm twice or Maleic hydrazide 500-ppm twice. These sprays can reduce the plant growth and does the job of pruning).

Pruning

In Maharashtra farmers have adopted the technique of manual as well as chemical pruning for giving rest to the plant and subsequently initiation of bahar. Even though no regular pruning is recommended, of late, pruning is practiced by some farmers to improve the fruit size. The fruiting in pomegranate takes place on one-year-old spur. The fruit that set on the terminal branches do not attain good size. To get such fruits, the terminal branches are pruned (15-20 cm) to avoid fruit set on them.

Pruned Pomegranate

Unpruned Plant

The systematic work on this aspect was initiated at CHES Vejalpur where manual pruning was carried out. Pruning in pomegranate was attempted to initiate new growth on 5 year old plants. The plants were pruned in the second half of May. Two types of pruning i.e. heavy (Removing around 5-7 kg fresh wt/plant and light (removing 2-3 kg of fresh wood/plant) was attempted. Under rainfed conditions. Sprouting started after 3-4 weeks of pruning. The subsequent flowering was however was delayed.

The mean fresh and dry wt of biomass removed in heavy pruning per plant was 5.67 and 2.92 kg/plant respectively, where as, in light pruning it was 2.90 and 1.65 kg/plant respectively. New shoot growth as evidenced by increase in shoot length, diameter, fresh and dry weight reveled that heavy pruning resulted in higher shoot growth compared to light pruning and control (no pruning) similar trend was recorded in leaf growth also.

Effect of pruning on new growth (fresh and dry wt of shoot)

Treat.	Shoot Fresh wt. (g) Month		Shoot Dry wt. (g) Month		Shoot length (cm) Month		Shoot Dia. (mm) Month		Pruned wood/plant	
	1	2	1	2	1	2	1	2	Fresh Wt. (kg)	Dry wt. (kg)
Heavy Pruning	6.37	10.04	3.16	4.75	66.33	70.83	4.51	4.98	5.67	2.90
Light Pruning	4.51	6.57	2.14	3.23	56.16	60.66	3.76	4.28	2.90	1.65
Control	1.58	2.49	0.70	1.22	26.83	30.5	2.42	2.95	-	-
CD 5%	0.51	0.73	0.29	0.47	8.50	9.04	0.26	0.23	-	-

Effect of pruning on new growth and nutrient composition

Treatment	Leaf Fresh Wt. (g) Month		Leaf Dry Wt. (g) Month		N% Month		P% Month		K% Month	
	1	2	1	2	1	2	1	2	1	2
Heavy pruning	8.78	11.27	4.07	4.61	2.94	2.42	0.18	0.13	0.850	0.712
Light pruning	6.12	8.21	2.70	4.02	2.91	1.52	0.16	0.13	0.698	0.66
Control	2.39	4.87	1.16	0.98	2.58	1.81	0.14	0.18	0.782	0.84
CD 5%	1.02	0.68	0.53	0.72	--	--	--	--	--	--

Effect of pruning on yield and yield parameters

Treatment	Fruit set no.	Fruit retention no./plant	% Retention	Yield kg /plant	Yield q /ha	Fruit wt g	Fruit length mm	Fruit Dia mm
Heavy pruning	67.8	52.0	85.54	12.7	50.0	169.33	65.21	71.33
Light pruning	75.7	63.5	88.75	10.4	41.6	158.33	58.88	61.05
Control	114.5	54.3	68.39	10.86	43.4	147.00	57.98	63.05
CD 5%	20.3	6.8	06.16	--	--	14.45	NS	7.72

Effect of pruning on NPK content of shoot

Treatment	N% Month		P% Month		K% Month	
	4	5	4	5	4	5
Heavy pruning	1.97	1.06	0.18	0.11	0.894	0.95
Light pruning	1.52	1.48	0.14	0.11	0.764	0.97
Control	1.81	1.52	0.18	0.13	0.834	0.77

Chapter 9

ORCHARD REJUVENATION

The old orchards become non-productive due to incidence of pest and diseases and neglect, once the orchard becomes old its productivity goes down, and it is general tendency of the farmers to neglect the orchard. In pomegranate when the survey was conducted it was observed that 25 percent of the old orchards become unproductive due to loss of vigor in the plant. Owing to heavy incidence of pest and diseases. There is still scope to review production from these orchards through intensive care. Rejuvenation can be bone by mechanical measures, chemical measures as well as by pruning of plants. Work carried out on non-selective pruning at CHES, Vejalpur has shown encouraging results. Fifteen-year-old orchard that was showing the signs of decline mainly due to drying of old branches resulting in sparse flowering and fruiting was used for the study. To over come the situation non-selective pruning of the orchard was undertaken. The plants were headed back to 30 cm, 60cm, 90cm,and 120 cm above ground level in the month of April, no pruning acted as a control. The cut ends were sprayed with chlorpyriphos 3ml/lit along with bavistin 1gm /lit to control the incidence of termites and various pest and diseases. New sprouts emerged after 20-25 days of pruning. Number of new sprouts per plant was reduced to 8 to 10, by allowing two to three healthy sprouts per branch to balance the framework of the plant. On the onset of monsoon the growth of the plant picked up and after one year the plant height was on par with that of control plants. In due course the plant growth slowed down due to diversification of food material to flowering and fruiting. Mean shoot diameter increased at a faster pace after the initial setback.

Fruit set in the initial years was reduced in the first year of pruning only 19 fruit /plant were set which surpassed the control in second year itself (101 no/plant) in plants pruned to 30 cm from ground level. It out performed control in the years to follow. Similar trend was recorded in respect of number of fruit retained / plant in other treatments. Maximum number of fruit were retained in plants pruned to 30 cm from ground level. Yield kg/plant was reduced in the first year after pruning (81.18% reduction) in treatment pruning to 30 cm from ground level. The yield/plant however has surpassed control in second year itself. The increase was to the tune of 16.96%, 32.94% and 26.10% in 2nd, 3rd and 4th year of pruning to 30 cm level.

Pruning plants to 30 cm from ground level was the best to rejuvenate the old pomegranate orchard and can be recommended as standard practice to rejuavation old orchards (Hiwale, 2007).

Growth parameters of Rejuvenated pomegranate cv. Ganesh

Treatment/year	Plant Height m		Stem Dia Mm		Plant spread			
					NS (m)		EW (m)	
	1999	2001	1999	2001	1999	2001	1999	2001
T1 Cutting 30 cm	2.68	2.42	21.23	30.94	2.39	2.55	2.43	2.73
T2 Cutting 60 cm	2.61	2.66	23.88	32.67	2.61	2.51	2.66	2.62
T3 Cutting 90 cm	2.68	3.34	17.18	34.03	3.46	2.72	3.46	2.72
T4 Cutting 120 cm	2.7	3.42	17.56	32.81	3.34	2.29	3.48	2.32
T5 Control	3.13	3.56	32.73	35.96	3.68	2.54	3.66	2.63
CD 5%	2.93	0.35	2.62	7.13	0.26	NS	0.35	NS

Hiwale (2008).

Fruit Set and retained /plant

Treatment/year	No. Fruit Set/plant				No. of Fruit Retained/plant			
	1999	2000	2001	2002	1999	2000	2001	2002
T1 Cutting 30 cm	19.0	101	89.2	53.00	7.6	52.4	60.8	51.8
T2 Cutting 60 cm	50.8	92.2	88.8	140.75	16.2	44.4	52.0	74.4
T3 Cutting 90 cm	63.4	70.8	82.2	42.75	20.0	35.4	38.4	34.6
T4 Cutting 120 cm	72.8	48	64.8	48.25	22.0	29.0	35.0	46.4
T5 Control	83.4	90.6	93.2	90.75	40.4	44.8	46.4	59.0
CD 5%	5.97	21.05	27.33	47.62	2.74	4.58	5.43	11.16

Hiwale (2008).

Percent Retention and Yield kg/plant

Treatment/year	% Retention				Yield kg/plant			
	1999	2000	2001	2002	1999	2000	2001	2002
T1 Cutting 30 cm	40.0	51.88	68.16	97.73	1.14	7.86	9.12	11.16
T2 Cutting 60 cm	31.89	48.16	58.56	52.86	2.43	6.66	7.80	7.77
T3 Cutting 90 cm	31.54	50.00	46.72	80.93	3.02	5.25	5.76	6.33
T4 Cutting 120 cm	30.22	60.42	54.01	96.16	3.10	4.35	5.46	6.96
T5 Control	48.44	49.45	49.78	65.01	6.06	6.72	6.86	8.85
CD 5%	-	-	-	-	0.39	0.69	0.92	0.48

Hiwale (2008).

Pomegranate Heading Back

Pomegranate Sprouting Back

Rejuvenated Pomegranate Plant

Fruit thinning

Fruit thinning is one of the tools in the hands of Horticulturist to reduce the burden of the plant and avoid the use of stored food material in one time. The sink source relationship plays important role in managing the crop load. Fruit thinning helps in maintaining the sustainability of production over the years particularly in fruit trees where, life span is long. By reducing the number of fruits the crop load can be evenly distributed all over the plant, there by-avoiding branch braking, better exposure of fruits to sunlight, which helps in better fruit development.

Effect of thinning of fruit on Yield and physico- chemical characters of pomegranate cv. Ganesh.

Thinning	Yield/ Plant (kg)	Fruit wt. (g)	Fruit Length mm	Fruit Dia. mm	Aril skin ratio	Juice (%)	TSS °Brix	Acidity (%)	Yield (Q/ha)	Gross income (Rs.)	Net return (Rs.)
25 fruit	4.61	197.74	66.26	65.86	1.66	74	17.08	1.82	18.44	7336	3726
50 fruit	8.95	185.82	63.22	61.89	1.77	73	16.16	1.84	35.80	14320	10670
75 fruit	11.41	155.62	61.74	60.02	1.87	73	16.0	1.79	45.64	13692	10042
Control	10.24	109.9	56.16	56.82	2.32	72	15.8	1.82	40.96	8192	4692
C.D. 5%	0.474	10.59	2.11	N.S.	0.01	NS	0.080	NS	-	-	-

(Hiwale, 2008).

Work carriedout at CHES Vejalpur, under semiarid rainfed conditions revealed that fruit weight showed maximum increase over control (79.92 percent) when 25 fruits/plant were retained; it was 69.08 percent with 50 fruit and 41.60 percent with 75 fruits per plant. However, there was no change in percent juice content and acidity of fruits. T.S.S. content was maximum when less number of fruits/tree were retained (17.08^0 Brix), showing a decreasing trend as number of fruits per plant increased. It was observed that keeping 75 fruits per plant produced maximum yield per hectare (45.64 Q) compared to control (40.96 G). Thinning fruit to 25 and 50 respectively numbers though resulted in reduction in yield by

54.98 percent and 12.56 percent. However fruit thinning produced good size fruit of 'B' grade, fetching better prices in the market. Overall keeping 50 fruits per plant was found to be economically viable preposition to the farmers as the economic returns were highest (Rs.10670/-) compared to control (Rs.4692/-), (Hiwale, 2008).

Rate of sale according to grade:	**Cost of cultivation** - Rs. 3500/-
B grade- Rs. 5/kg- Rs. 400/q	**Cost of thinning** - 3 labor Rs./ha @Rs.50=Rs. 150/-
C grade- Rs. 3/kg- Rs. 300/q	**Small Fruits** - Rs. 2/kg- Rs. 200/q.

Chapter 10

CROP REGULATION, ROOT DISTRIBUTION AND INTERCROPPING

Crop Regulation

Staggering in pomegranate is practiced with a view to regulate crop load on the plant which will help in availability of the fruit for the long period helping the farmer fetching remunerative prices in the market. Also it will help in avoiding the glut in the market leading to distress sale. The relationship between source and sink can be regulated which helps in better fruit development. Pomegranate flowers and fruits through out the year if irrigation is provided to the plant. Taking of particular bahar leads to the glut in the market. This results in distress sale by the farmers. To avoid this staggering of pomegranate crop was attempted to find out and fruiting flowering during various months of the year so that maximum production with best quality fruits can be obtained. The pomegranate flowering was allowed during different months of the year in five year old plants of cultivar Ganesh by withholding irrigation and releasing it during the specified period to initiate flowering. Best crop was obtained when September flowering was allowed because the fruit development coincided with maximum moisture availability and cool climate leading to less incidence of insect pest and reducing cracking leading to improved quality of fruits.

Staggering in Pomegranate

Month	Fruit length (mm)	Fruit Dia. (mm)	Fruit wt. (g)	Yield/ plant (kg)	Aril seed ratio	100 seed Wt (g)	T.S.S.	Acidity %
August	67.4	78.10	134.20	9.83	1.89	1.54	16.18	0.44
September	70.99	70.24	151.68	11.83	2.70	1.89	17.64	0.35
October	70.92	70.70	141.88	9.98	3.03	1.96	16.72	0.37
November	70.75	71.05	141.66	2.88	3.28	2.36	18.12	0.34
C.D. 5%	2.11	N.S.	2.91	2.35	2.51	0.17	1.03	0.07

It was observed that in September flowering produced maximum yield per plant. Aril seed ratio increased and softness of seed decreased with the period. Time of flowering significantly influenced T.S.S. and acidity. T.S.S. increased with increase in time of flowering while acidity showed decline. The best fruits were obtained when September flowering under rainfed conditions with one or two supplementary irrigations. Similar observations were recorded by Sonwane and Desai (1989).

Chemical Regulation

Fruit and leaf abscission have been reported with application of Alar at 500-3000 ppm. However, now a days, chemical regulation is avoided as most of the chemical used for it have been banned, are reported to be harmful or are not manufactured.

Pomegranate Hasta Bahar

Pomegranate Ganesh Mrig Bahar Crop

Bahar Regulation

The main purpose of crop regulation is to force a tree into rest and to induce profuse flowering during the season, like in pomegranate there are three flushes of flowering i.e. Mrig (June-July), Hasta (September-October) and Ambe-bahar (December-January). The methods used for imposing bahar can be broadly classified into cultural and chemical. In cultural method the common approach is to withhold irrigation, root exposure, root pruning etc., are used. In chemical methods most of the growth retardants and growth regulators can be used to impose the bahar treatment. In pomegranate flowering is natural in Mrig bahar, which depends upon monsoon and is difficult to regulate. But withholding irrigation can regulate Hasta and Ambe bahar. Thus imposition of stress leads to leaf drop and cessation of growth. Later on reducing the stress slowly will lead to profuse flowering and fruit set. At CHES, Vejalpur, Raturi and Hiwale (1991) working on pomegranate cv. Ganesh under rain fed condition reported that it is possible to raise successful Hasta bahar crop. Hand thinning was used to remove Mrig bahar flowering. Comparative studies on Mrig and Hasta bahar indicated the fruit set, retention and yield per plant were highest in Hasta bahar. The fruits of the bahar are available when there is no glut of pomegranate in market and hence fetch remunerative price. The quality of fruit in respect of skin color, T.S.S. was also superior. Also the incidence of pest and diseases was also least. Fruit borer infection in Mrig bahar was recorded up to 18.83 per cent whereas, in Hasta bahar it was just 8.92 per cent.

Effect of cropping season on fruit retentation and yield of pomegranate cv. Ganesh

Cropping season	% Fruit set		%Fruit retention		Yield/plant (kg)	
(Bahar)	88-89	89-90	88-89	89-90	88-89	89-90
Mrig	3.90	3.99	39.97	47.63	3.18	5.16
Mrig+ Hasta	5.04	5.43	42.49	55.46	4.52	7.37
Hasta	8.12	9.26	54.06	65.81	5.82	10.25
C.D. at 5%	1.26	0.79	9.43	5.55	1.01	2.18

Raturi and Hiwale (1991)

Improving Fruit Skin Color

Normally when temperature is high during ripening period, the skin color of Ganesh variety becomes dull and pale. The fruit losses market value. When there is excess nitrogen, irrigation and magnesium in the soil, the skin color becomes light and dull. Therefore, when fruits are maturing and turning their colour, potassium application helps in improving the fruit color. The color of aril and skin are due to anthocyanin pigment (Due *et al.* 1975). Low temperature stimulates anthocyanin synthesis and activity of enzyme phenyl aniline ammonia lyase (PAL), which is directly related to anthocyanin synthesis (Akoi et al, 1970 and Hyodo, 1971). Spray of potassium dihydro orthophosphate 1-2 g in 1 liter of water or two sprays of 500 ppm Lihosin. Also two sprays of 500 ppm of ethrel during color breaking stage along with cytozyme 100 ml + 100 g potassium dihydro orthophosphate in 100 liters of water improved skin colour.

Root Distribution

Root distribution acts as guide for application of fertilizers as well as irrigation. Root distribution varies according to the type of soil and method of propagation. Studies were initiated to findout root distribution patter of air layered six year old pomegranate cv. Ganesh. Root distribution pattern in air layers of pomegranate cv. Ganesh under semiarid rainfed conditions revealed that the root system is shallow in nature and within the plant canopy as below 60 cm soil depth not much root activity was recorded. Maximum root activity on fresh and dry wt. basis was observed at 0-50 cm radial distance from tree trunk (54.17%), which decreased to 31.12% at 50-100 cm radial distance and 2.12 % at 100-150 cm radial distance. However soil depth wise it was maximum (53.55 %) at 30-60 cm soil depth and 42.69% at 0-30cm depth. It was least (3.74%) at 60-90 cm depth. Similar observation were recorded in respect of Ber by Hiwale (2002).

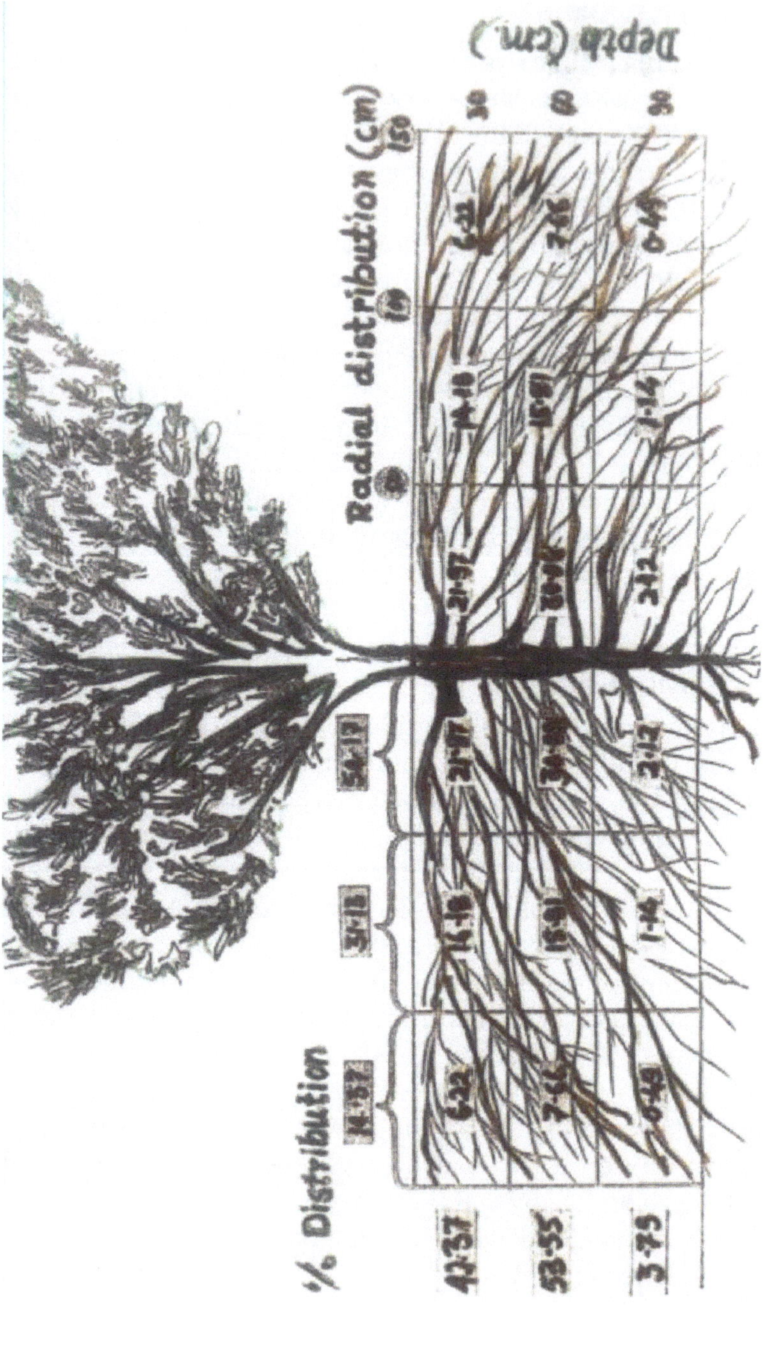

Spatial Root Distribution in Pomegranate cv. Ganesh (Perce)

Root distribution Pattern (fresh and dry wt. g/plant)

Type of root	Depth of soil	0-50 Fresh wt.	Dry wt.	50-100 Fresh wt.	Dry wt.	100-150 Fresh wt.	Dry wt.	Total Fresh wt.	Dry wt.
Thin (>3mm)	30cm	190	114	250	173	101	67	541	354
Medium(>7mm)	30cm	282	162	185	122	90	50	557	334
Thick (>15mm)	30cm	202	118	-	-	-	-	202	118
Total		674 (21.97)	394	435 (14.18)	295	191 (6.22)	117	1310 (42.69)	806
Thin (>3mm)	60cm	272	213	216	153	138	100	526	466
Medium(>7mm)	60cm	364	304	164	101	97	73	625	478
Thick (>15mm)	60cm	287	228	105	40	-	-	392	268
Total		923 (30.08)	745	485 (15.81)	294	235 (7.66)	173	1643 (53.55)	1212
Thin (>3mm)	90cm	--	--	--	--	--	--	--	--
Medium(>7mm)	90cm	--	--	--	--	--	--	--	--
Thick (>15mm)	90cm	65	37	35	21	15	9	115	67
Total		65 (2.12)	37	35 (1.14)	21	15 (0.49)	9	115 (3.74)	67
Grand Total		1662 (54.17)	1176	955 (31.12)	610	441 (14.37)	299	3068	2085

Hiwale *et al.* (2009).
(Figures in parentheses are percent values)

Percent root activity was calculated as suggested by Bhojappa and Singh (1975)

$$\% \text{ Root distribution} = \frac{\text{Root wt. at particular location}}{\text{Total root wt. at all depths}} \times 100$$

Intercropping

Pomegranate occupies the recommended spacing (5m x 5m) after 8 years, but the trees start bearing at 4 years age and until then inter-cropping can be done with fodders like berseem and Lucerne and Leguminous crops like cowpea and green gram and vegetables like cucurbits, cabbage, cauliflower, beans, peas, tomato, carrot, onion, radish, potato, brinjal etc. However inter crops should be regulated so that their cultural care does not interfere with the bahar regulation. Similarly the intercrop should not help in pest carryover leading to heavy past incidance on the main crop.

Chapter 11

POST HARVEST MANAGEMENT

Fruit Maturity Criteria

1. The fruit takes about 135 to 170 days to mature depending on the variety.
2. In summer, mature fruit turns yellow and is dark pink in winter and rainy season.
3. The calyx at the bottom of the fruit turns hard and curves inwards.
4. The fruit shape gets compact
5. The fruit skin becomes hard.
6. The color of the aril turns dark red or pink and is full of juice and sweetness.

Fruit Harvesting

There is great demand for pomegranate at national as well as international market and there is a scope to earn the valuable foreign exchange through its export. This increases the need for utmost care in harvesting the fruit at right maturity, packing and storage. Ambe bahar fruits are ready for harvest in June to August, where as, Mrig bahar fruits can be harvested in November - December. Hasta bahar fruits in February -March. The fruits should not be retained on plant for long time as it results in reduction of fruit quality. The aril turns brown and taste of the fruit is lost. Fruit harvesting should be done with the help of a secature and not by twisting it as the fruits are born in clusters of 2-3. The fruits are ready for harvest in 4 ½ to 5 months after the appearance of blossom. The fruits are harvested, when the skin turns slightly yellow and the fruit gives metallic sound when tapped or pressed. The tree may give 100-150 fruits per year and economic yield can be obtained up to the age of 20-25 years.

Fruit Grading

Fruit grading is important for export as well as for local markets to obtain reasonable price. Cracked, split, diseased and borer-infested fruits should be separated out and pre cooling of the fruits at farm level should be carried out preferably in cool chambers or either in the shade of a plant. The fruits should be graded according to maturity, weight, size and colour in to following four grades.

A. Attractive, Dark red, bigger size fruits having fruit weight more than 350 g, without any blemishes are classified in this grade.

B. Big fruits weighing around 300-350 g with out blemishes and attractive color.

C. This is the third category of fruits weighing around 200 to 250g without blemishes and attractive color.

D. Fruits with 200 and below weight without blemishes and attractive color can be categorized in this grade.

Packaging

Right type of packaging is advantageous because :

- Transportation of fruit to distant markets becomes easy.
- Handling or shifting material at various steps becomes convenient.
- Attractive packing definitely attracts the dealer as well as consumer.

The fruits are packed according to grade, size of box. The size of box will vary according to grade. There are 4 fruits of grade A in a cardboard box and six fruits of grade B in a box of 13″ length and 9″ breadth and 4″ height. Nine fruits of grade C in a box of size 15″ length and 11″ breadth with 4″ height. Twelve fruits of grade D in a box of 14″ length and 10″ breadth with 4″ height. After grading, these fruits are sometimes dipped in ethyl oleate to provide luster. For export markets, the fruits are wrapped in tissue paper and arranged in two rows in 3 ply CFG boxes. Boxes made of lightwood are also used. Dry grass is used at the bottom and top of the box or baskets to set as a packaging material. Up to 6 dozen fruits are packed in each basket and transported by trucks.

Storage

Pomegranate keeps well for a long time. Experiment showed that fruits stored at 0° C and 4.5°C at 80-85 percent relative humidity did not undergo any shrinkage or spoilage for seven months. Studies at Indian Institute of Horticultural Research, Bangalore, showed that shelf life of pomegranate cv. Ganesh was 3 weeks at 25°C and 10 weeks at 8° C storage. Film wrapping retained the quality, surface colour and reduces moisture loss. Shrink-wrapping of pomegranate extended the shelf life to 70 days at 8°C as compared to 20 days at 25°C storage. For bulk storage of fruits, they are packed in layers in wooden crates containing 16-18 kg of fruits and stored at 0°C and 80 percent relative humidity.

Post Harvest Technology

The pomegranate had a low respiration rate and a non-climacteric respiratory pattern. Freshly harvested pomegranate fruits of cv. Ganesh were treated with wax coupled with carbendazim (0.1 %) and stored at room temperature (22-24^0 C with 52-82 % RH) as well as in cold storage. Results indicated that fruits could be stored up to 75 days in cold storage as against for 30 days at room temperature (Wasker *et.al*, 1999) Chilling injury can be overcome by intermittent warming. Storing pomegranate at 2^0 C with intermittent warming was found optimum for minimizing chilling injury and maintaining the fruit quality up to 13 weeks (Artes *et al*, 2000). Controlled atmospheric storage of pomegranate fruits indicated that the color value of the juice was significantly increased in fruit stored at 10 percent $O_{2,}$ 5 percent $CO_{2.}$ (Artes *et. al.*, 1996). The pomegranate fruit reached its horticultural maturity when its soluble solid content reached 15% (Arie sengal and Reich, 1984). The maturity period of 105-140 days from fruit set can be used as one of the criteria for fruit maturity (Heys, 1953). However, Kulkarni et.al. (2005) reported that harvesting of fruit at 100th day from fruit set was found to be crucial to prevent browning during storage.

Value Addition

With the increase in area and production of the fruit, if steps are not taken to develop industry for value addition of a crop, results in glut in the market and the crop becomes uneconomical, leading to either distress sale or removal of orchards. Pomegranate fruit is known for its value added products like juice, jelly, anardana, anar rub, rind power etc. can be prepared at the field level with ease. Fruits are mainly used as fresh fruit or juice consumption.

Fruit Juice

The varieties having dark red aril color are preferred. The juice can be extracted by two methods. Keeping the skin intact and removing the lower portion and the fruit is separated in to four parts and the juice is extracted. In the second method the skin is removed, arils are separated and juice is extracted. Adding potassium metabysulphite 2 percent and storing it in cold temperature in steel or glass bottles/vessels preserves the juice. For pasteurization of fruit juice a temperature of 77^0 C for 33 seconds is recommended. The process eliminated *Aspergilus niger* and lactic strains of Bacteria. Colour stability was maintained up to 80^0C temperatures for 30 min. After which there was slight color reduction. Pasteurized juice showed comparable quality to unpasturized juices (Rage and Pai, 1999). The pomegranates Juice freezed at - 40 ^{0}C either with or without sugar and stored at -18^0 C for 9 months. (Bilsil and Cevic, 1997). Clarification of pomegranate juice by ultra filtration showed better retention of ascorbic acid (2-5 % loss, high yield (97.8%) and better quality (98.55) than conventionally clarified juices (Iboyaima et al 1993). Effective stabilization of color changes in juice stored at room temperature can be done with polyvinylpyrrolidine (PVP) and trypsin (Khareeva *et al.,* 1987).

Pomegranate Syrup

The product is prepared by heating fruit juice and raising the sugar to 60-64 percent. The syrup can be stored under normal conditions for about 3-4 months. The syrup is served by mixing water in appropriate proportion according to the taste.

Pomegranate Nectar

The fresh pomegranate juice can be preserved for long time by increasing the sugar content to 20-30 percent.

Juice Concentrate

Juice is made in to powder form either by freez drying or by drying on conveyer belt and is packed by vacuum packing in polythene bags.

Pomegranate Chutney

Wild pomegranate (Anardana type) found in Himachal pradesh can be used to prepare chutney. The product was dried in a mechanical cabinet at 65^0 C for 10-14 hours. Samples were packed in polythene pouches and stored at 37^0 C. The quantity of green chillies added increased the vitamin C content, total sugars and pH, while protein levels and acidity decreased. Chutney had a shelf life of more than six months.

Carbonated Pomegranate Beverages

The pomegranate beverages were prepared by different levels of juice (5%, 10%, 15%, 20%). Drinks were formulated to pH of 3.14-3.16, acidity of 0.24 % and stored for 80 days at ambient temperatures. The beverage with 10 percent pomegranate juice with initial content of 11.78 percent sucrose, 1.3 percent reducing sugar and 1.0 mg ascorbic acid was the best (Riaz and Elahi 1992).

Pomegranate Wine

Pomegranate wine is one of the important products where it is possible to absorb the glut of this fruit in the market, simultaneously providing healthy drink to the people and creating job opportunities. The pomegranate fruit of one kg yields around 500 ml of juice from which 400-450 ml wine can be prepared through fermentation with addition of yeast.

Methodology

Preparation of Starter- A spoonful of standard wine yeast *Saccharomyses ellipsoids* no.101 maintained on yeast extract peptone dextrose agar (YEPD) was transferred to a test tube containing 5 ml of YEPD broth and incubated over night at

25°C., after 22 hrs. it is shaken well in a incubator at 120 rpm which is used for inoculation after activating the yeast cell at 40°C for 1.5 hrs.

Steps in Wine Making

Pomegranate fruits
↓
Washing
↓
Aril separation
↓
Juice Extraction
↓
Add potassium metabisulphite (200mg/lit of juice)
↓
Raise the Brix of the juice to 24°
↓
Addition of wine yeast culture (5%)
↓
Fermentation at 26°c for 9 days
↓
Separation of wine by cheese cloth
↓
Filtration
↓
Sealing of filled bottles
↓
Pasteurization at 65°C for 20 minutes
↓
Storage of wine

Production Cost of Pomegranate Wine

Particulars	Cost (Rs.)
Pomegranate Fruit (1Kg)	10.00
Juice extraction cost (500ml juice / kg)	0.50
Preservative cost	1.00
Processing cost	5.00
Cost of one liter wine	40.00

Anardana

Wild pomegranate arils are sun dried and commercially marketed as "anardana", which is used as condiment (Sharma and Sharma, 1990). Chemical characters such as high juice content and sugar/acid ratio, more intensive color and less tannin content are good attributes for a better quality of anardana. To improve color stability by preventing oxidation of the anthocyanin, a mixture of glucose and citric acid is used as stabilizer at a ratio of 1.0:0.20: 0.003 of juice to glucose to citric acid (Gafizov *et al,* 1990).

In Himachal pradesh this is a household activity and anardana is mainly used as substitute for tamarind and mango slices. Also used in garnishing of ice cream, fruit salad, chutney and pan-masala. There is huge demand for the product in North India. The culled fruits can be used for preparation of anardana. Anardana can be prepared from varieties having 7.8 to 15.4 percent acidity specially the Russian types are more suitable. The varieties under large-scale cultivation are having acidity of just 0.3 to 0.5 percent, which can be used by adding five percent citric acid to the arils. The total expenditure for preparation of one kg anardana is around Rs.60/kg.

Steps in Anardana Making

Washing of fruits

↓

Extraction of arils

↓

Addition of 5% citric acid to increase acidity to desired level

↓

Drying in solar drier or open sun

↓

Packing

Export of anardana from India (Pruthi 2001)

Importing country	Quantity (Kg)	Value (Rs.)	Importing country	Quantity (Kg)	Value (Rs.)
USA	3615	2,53,888	Kuwait	100	6,723
Canada	2259	1,06,331	Australia	50	3,752
Sweden	184	44,342	Hong Kong	38	3,546
Newziland	575	44310	Malaysia	25	1,782
Germany	240	19544	Singapore	10	1,112
UAE	220	11364	Oman	50	938

Mahajan *et al.* (1992) processed wild pomegranate for preparation of anardana with three thermal treatments viz. deep sand roasting of hole fruits for five minutes, dry roasting on heater and blanching in boiling water for five minutes .The first one was found to be the best. Three drying techniques viz solar drying, conventional sun drying and hot air dehydrator. Solar drying was the best. Hussein *et al.* (2004) analyzed pomegranate whole and seed powder samples and suggested standard grades for sensory evaluation as follows :

Proposed specification for pomegranate seed (whole)

Characteristics	Limits	
	Standard Grade	Special Grade
Organic Foreign matter % by mass, maximum	2.0	1.0
Inorganic Foreign matter % by mass, maximum	0.50	Nil
Damaged and discolored seeds % by mass, maximum	4.0	2.0
Moisture content % by mass, maximum	14.0	12.0
Total titratable acidity (expressed as Citric acid)	2.0	12.0

Proposed specification for pomegranate seed (power)

Characteristics	Limits	
	Standard Grade	Special Grade
Moisture content % by mass, maximum	12.0	10.0
Total Ash % by mass, on dry basis, maximum	6.0	3.5
Acid Insoluble Ash % by mass, on dry basis, minimum	0.50	0.30
Non-volatile Ether Extract % by mass, on dry basis, maximum	8.0	12.0
Crude Fiber % by mass, on dry basis, maximum	22.0	18.0

Problems in Anardana Preparation, Storage and Marketing

- The entire operation of collection of fruit, separation of seeds from rind and drying of seeds is done manually involving intensive labor.
- Uneven size and unequal maturity period of fruits arising out of unknown varietal sources without quality grading, often poses problems in uniform drying and thereby to produce uniform product.
- Drying of pulp-coated seeds in open sun over a long period causes contamination with dust and dirt.
- Some times fungal growth occurs due to intermittent rain or unfavorable weather. It may result in caking of the product.
- Lack of appropriate packaging decreases the quality and may invite pest and disease infestation.
- Color and flavor is often reduced due to bad storage condition.

Chapter 12

DISORDERS

Fruit Cracking

The results fruit cracking is observed in almost all the cultivars. The extent of cracking ranged from 3.57% in kazkai to 76 .6% in Suni Bedana, Josan *et al* (1979). Fruit cracking is mainly associated with fluctuation in soil moisture, day and night temperature and relative humidity and rind pliability and nutrient imbalance. Phadnis (1974) reported that irregularity in watering and lack of moisture at the time of fruit development leads to cracking. Singh *et al* (1967) reported that cracking might be due to boron deficiency as well as wide variation in moisture content of the soil and humidity in the air. Prasad and Bankar (2000) also reported that varietal character plays important role in cracking. The percentage was minimum (36.6%) in Jalore seedless and maximum (78.2%) in Jodhpur red.

Under semiarid conditions also the disorder is very much prevalent in pomegranate orchard and about 10-35 percent fruits are cracked depending on the variety, season and moisture availability in the soil. Most of the commercial cultivars are susceptible to this disorder. Hiwale and Raturi (1999) reported that incidence of cracking was dependant on the cropping season. Cracking is more in Mrig bahar fruits because of variation in humidity, but this can be reduced by regular irrigation during rainless periods. Fruit cracking was maximum in Mrig bahar crop (30-35%) followed by Ambe bahar (25-30%) and was least in hasta bahar crop (10-15%).

Randhawa *et al.* (1988) reported that cracking occures after a heavy rainfall followed by drought.

Prolonged dry spells cause hardening of peel. If rains follow dryness, then the pulp expand and the peel cracks. Sometimes thrips and mites attack on rind, where they suck the sap also causes fruit to crack. Imbalance in nutrition i.e., excess use of nitrogen may lead to cracking. The disorder may also be due to Boron deficiency in young fruits and moisture imbalance in mature ones. When growth hormones are used with excess or less concentration also causes fruit cracking. Irrigation at 75 percent field capacity along with application of 0.6 kg nitrogen per tree has reduced cracking (Singh *et al.* 1993). During fruit growing period, a spray of 1 kg calcium chloride + 1 kg magnesium chloride dissolved in 100 liters of water can reduce fruit cracking. This can be followed by another spray of 1.5 kg DAP +0.75 kg magnesium sulphate in 100 liters of water. Further if needed, another spray of 90 ml cytozyme along with 100 g borax in 100 liters of water is given.

The work carried out to control the disorder indicated that spraying of micronutrients like boron at 0.5%, calcium at 1 percent and iron at 1 percent are found to reduce the cracking by 20-25 percent (Sheikh and Patil 2005). Bohlmann (1962) opined that application of boron checks cracking of apple fruit, which was due to role of boron in synthesis of pectic substances in the plant. Misra and Khan (1981) suggested that boron helps in translocation of sugars and synthesis of cell wall due to the activity of enzyme methyl esterase.

Internal Breakdown

The problem is noticed to a great extent in all the recommended varieties. This is also known as blackening or browning of arils. Out of the eight locules, single locule is affected. Externally the fruit looks healthy. The melody was reported as early as 1974 by Ryall and Pentzer (1974) They reported that arils did not develop proper color and had streaked appearance due to white lines radiating from seeds and arils posed abnormal flavor. The causes and remedies of this malady are not known exactly. The problem seems to be related to various aspects like variety, stage of harvest, bahar (Season), location of fruit, biochemical changes in fruit during the process of ripening and pathological causes. Prabhu Desai (1989) and Khodade (1987) reported that selection P-23 at MPKV Rahuri, had low percent of affected fruit than Ganesh. Shete (1998) revealed that the incidence is high in Mrig and Ambe bahar in different varieties. Highest incidence (100%)was recorded in G-137 and Jalore seedless from 105 days onwards during mrig bahar. Among all the cultivars Mridula showed minimum incidence (50%) at 120 days after fruit set, later on it reached to 100 percent. Stage of maturity at harvest, delayed harvesting of fruit beyond 150 das might be the possible reasons of browning, (Khodade, 1987). Increased pruning intensity up to 60 cm of main stem with thinning resulted in reduced breakdown, (Pawar, 1993). Fruit location on the tree also influenced the incidence of breakdown. Prabhu Desai (1989) reported that the incidence was maximum on southern side of tree than the fruits in the center. Nutrition also plays an important role, Prabhu- Desai (1989) recorded significantly higher amount of Nitrogen, Potassium, magnesium and boron in infected than the healthy arils. Spraying of ascorbic acid, on the pomegranate plants as an antioxidant showed significant effect on all biochemical parameters. Shete (1988) and Sawant (1993) observed low calcium content in the affected fruits. No specific relation was observed in iron and copper content of healthy and affected fruits. Blackening of fruits has also influenced the chemical characters of fruit. Prabhu Desai (1989), Sawant (1993), Shete (1988) reported affected arils were low in TSS, ascorbic acid, acidity total sugar, reducing sugar. The tannin content of the affected arils was high and the increase ranged from 32 to 93.61 percent. Enzyme activity of polyphenol oxides and peroxides increased in affected arils (Prabhu-Desai, 1989, Sawant, 1993, and Shete, 1988). The activity of catalase and alpha amylase was significantly reduced in affected arils (Prabhu-Desai, 1989, Sawant, 1993).

Chapter 13

EXPORT

Export

India is one of the largest producer of the pomegranate. Apart from India, Spain, Iran, Egypt, Peru, Israel, Pakistan, Afganisthan, Turkey and America are the countries where the crop is cultivated. World production of Pomegranate is around 10 Lakh tones out of which India's share is 40-45 percent .At present Spain is the maximum producer and exporter of pomegranate due to high quality and low transport cost capturing most of the European market. The total demand for pomegranate in the world is around 1.5 Lakh tones, which is increasing day by day. Spain contributes 75,000 to 80,000 tones (75%)of the total demand followed by Iran 25,000 tones and India 10000 tones (5%). Thus to increase the export there is only one way to improve the quality. The pomegranate export from India, which was 1790 tones in 1991-92, increased four fold to 4144 tones during 1994-95 and by the year 1999-2000 it was 5726 tones. There was decline in export there after and no definite trend was recorded. In 1991 to 1997 pomegranates were exported at the rate of Rs. 12 to15 per kg, which increased to Rs. 22 in 1999-2000. Pomegranate was mostly exported to United Arab Emirates (33 to 67.54 percent) Bangladesh (1.23 to 33.81 percent), Saudi Arabia (3.60 to 22.82 percent). After 1999-2000 there was reduction in export to Saudi Arabia and Bangladesh and export to countries like Bahrain, Kuwait, Oman, Srilanka and England increased. Export to countries like Qatar, Canada Germany, South Africa and Switzerland was around one percent. It was less than one percent to Singapore, France and USA, which are the major consumers. Therefore, there is need to survey these markets, define the requirements of the consumers so that export to these countries can be increased.

Export specifications

Variety	Middle east	Netherlands/ Germany	U.K.
Ganesh	300-450 g (red)	250-300 g (red)	250-300 g (red)
Arakta, Mridula, Bhagwa	200-250g Deep red	200-250 g Deep red	200-250 g Deep red
Packing	5kg	3kg	3kg
Storage	5° C	5° C	5° C

Export of pomegranate

Year	Export (tones)	Cost (Rs. Lakhs)	Year	Export (tones)	Cost (Rs. Lakhs)
1991-1992	1790	214.8	1998-1999	4239	938.09
1992-1993	1624	211.12	1999-2000	5726	1153.22
1993-1994	2623	367.22	2000-2001	4455	1022.37
1994-1995	4114	555.39	2001-2002	4,773	1041.85
1995-1996	4255	653.14	2002-2003	6,303	1434.56
1996-1997	4768	667.52	2003-2004	10,315	2109.00
1997-1998	5600	860.72	--	--	--

Bottlenecks in export are, high cost of transportation and low quality of produce. The main variety Ganesh is having unattractive aril color. Variety Arakta though the aril color is attractive it is less sweet .At present there is very good demand for fruits of a Bhagva as both the lacunas in Ganesh and Arakta has been overcome. Though there is export demand, due to fluctuation in rate it becomes impossible to maintain steady supply.

There is increasing demand for implementation of Euro gap certification from European countries. Stringent European standards for export and limited infrastructure facilities for export limits the export.

Supply from Spain is mostly during the month of October to December, where as pomegranate is available through out the year in India. Advantage of this fact should be taken to increase the export. There is scope to enter in to Russian market. There is a need to develop technology so that pomegranate arils can be sold and encourage farmers to obtain Euro gap certificates to increase export to Europe. There is a scope to export pomegranate to countries like China, Japan Australia and African countries by convincing the people, its medicinal value.

Export of value added products like Anardana, Squash, Syrup etc. along with processing technology of arils for export of ready to eat packing in European and American markets needs to be developed to reduce transportation costs and better utilization of available infrastructure. There is a glut in the market of pomegranate during June to August and the rates go down. Pomegranate if stored in cold storage its storage can be prolonged up to 60-75 days. Thus, extend it is possible to the period of availability of the fruit for export and halt the falling prices.

Availability chart of Pomegranate in the World

	April	May	June	July	August	Sept.	Octo.	Nov.	Dec.	Jan.	Feb.	Mar.
Spain						✓	✓	✓	✓	✓	✓	✓
India	✓	✓	✓	✓	✓	✓	✓	✓	✓	✓	✓	✓
Egypt					✓	✓	✓					
Iran	✓			✓	✓	✓	✓	✓	✓	✓	✓	✓
Peru	✓	✓	✓	✓								
Cyprus						✓	✓					

Post Harvest Handling of Pomegranate Fruits for Export

Selection of Pomegranate Orchard

Last pesticide application should be at least one month before harvest

Therefore should be healthy, free of nutrient deficiencies, sunscald, fruits clean not strewn with fungal/bacterial infected leaves and free from fruit borer infested fruits.

Fruits size should be uniform characteristic of the variety.

Fruits reached should have maturity and colour, with attractive and appealing skin color.

Maturity and Maturity Standards

Distinctive fruit colour characteristics of the variety. Metallic sound on tapping.

Ganesh cultivar - TSS reaches 14.5° Brix or acidity 0.3 to 0.35 percent.

Mridula cultivar - TSS-16° Brix, Acidity-0.45 percent.

Quality Requirements of Importing Countries

Smooth cutting at the stem end, Shining dark rose colour of the fruit

- Fruit weight around 200-400g
- Round shape of fruit
- Uniform size and shape of fruit in a pack
- Dark rose pink arils
- Soft seeded fruits
- High total soluble solids near about 16° Brix to 17° Brix
- Freedom from scars, blemishes, scratches, rusting, rough surface, dieses spots and fruit borer infestation.
- Calyx fresh should be and without damage.
- Pleasant flavor and aroma
- Without blackening of arils.

Method and Time of Harvesting

1. Harvest pomegranates early in the morning or in evening hours.
2. Use clippers and bags for harvesting.
3. Do not pull the fruits.
4. Cut the stem end close to the base.
5. Do not bruise the base while cutting the fruits.
6. Do not keep longer stem end, as it can cause damage to other fruits.
7. Keep the fruits in plastic crates.
8. Do not keep fruits on the ground.
9. Do not keep crates in the sun but in shade.
10. Do not overfill the crates with fruits and keep them in shade and not in open sun.
11. Do not damage calyx while harvesting or while keeping in crates or during field storage.

Receipt of Pomegranate at Pack-House

1. Unloading should be done to avoid
2. Unloading should be preferably in sorting bay having cool temperature
3. Avoid multiple handling of fruits
4. Do not stack too many crates over each other
5. Note down both tare and net weight of crates.
6. Growers code is to be maintained to avoid mixing.

Sorting, Washing and Treatment with Fungicide and Wax on Packing Line

1. Sorting

Inspection on roller-cum-conveyer. Fruits showing greenish colour, fruits with sunspots, affected by fruit borer, disease spot, rusting, rough surface to be eliminated. Having cuts, blemishes, damages must be rejected.

2. Washing

- Use of disinfectant-sodium hypochlorite @ 100-200ppm.
- Rinsing with clean water
- Potable water should be used

3. Waxing and Fungicide Application

- Only good quality edible wax must be used
- Monitor edible wax quantity
- Monitor concentration of the fungicide to avoid crossing of international residue limit
- Effective wax coating is very important

4. Drying the Wax Coating

- Temperature of drying tunnel must not to beyond 30-35^0 C
- Bulbs, heaters, hot air circulation along with fans, air kniver must work properly for proper drying

Sizing of Fruits

S. No.	Different size ranges	No. of fruits in 4.0 kg box	No. of fruits in 5.0 kg box
1.	450 to 500	Not exported (Indian Domestic market)	
2.	400 to 450		
3.	350 to 400	10-12	12-15
4.	300 to 350	12-15	15-18
5.	250 to 300	12-15	15-18
6.	200 to 250	15-18	18-20

1. For export purpose mostly 200-400 gm fruit weight are used
2. Variation in a size, grade should be minimum.
3. Bad quality of fruits should not be mixed with good quality fruits.

Nose Cleaning

Before packing a box, fruits at stylar end or calyx are wiped with a cloth soaked in a fungicide solution (benomyl or Thiabendazole 10000 ppm) to avoid cross contamination in a packed box.

Packing and Coding

1. 4.0 to 5.0 kg box is used with 10-12 number of fruits per box.
2. 5.0 kg box containing 10-15 fruits of Ganesh and Mridula or Arakta 18-20 numbers
3. Dimensions of 4.0 kg box –375 x 275 x 100 mm and 5.0 kg- 480 x 300 x 100 mm

Characteristics of Boxes

1. Cardboard boxes must have ventilation holes, hand holes etc.
2. Fruits should not press each other to avoid bruise on the skin.

3. No movement of fruits inside the pack by putting accurate size of fruits.

Pallets should be 1000mm x 1200 mm. For export to Europe, dimension of 800mm x 1200 should be used. Only plastic material should be used for strapping and clipping for securing the cartons to the pallets.

Information on Boxes

1. Name of produce
2. Variety
3. Origin of produce
4. Commercial specifications of class
5. Date of packing
6. Name of packing organization
7. Gross weight/net weight
8. Number of fruits/fruit size
9. Recommended storage temperature
10. Additional labeling requirement of importer should be included.

Pre-Cooling

1. Before storing, the fruit needs to be pre- cooled by forced air circulation system.
2. The ideal pre-cooling temperatures must be 5-7^0 C at least within 10 hours after harvesting.
3. Chilling damage of fruit is possible if stored below 5^0 C.
4. Humidity should be around 90-95% during pre-cooling.
5. Prompt pre-cooling is very important for avoiding post harvest losses.
6. Filled cartons should be shifted to pre-cooling chamber without any delay.
7. On transfer of product, outflow temperature and humidity should be properly recorded.

8. Product should remain in the pre-cooling chamber till the temperature of the fruits reaches to required stage.
9. Do not allow the temperature, to fall otherwise chilling injury will make the fruits unfit for consumption.
10. Temperature, humidity and air circulation of the room shall be properly maintained
11. After pre-cooling and palletisation, strapping should be done. This needs to be done only in cold room to maintain the cool chain.

Cold Storage

1. The ideal storage temperatures are 5-7^{0}C
2. Relative humidity should be kept at 90—98% throughout transportation and storage.
3. Chilling also varies with maturity and variety of Pomegranate develops brown discoloration of skin, pale colour of arils and brown discoloration of white segments separating the arils.

Quality Appraisal

Quality checking should be done with regard to following parameters.

1. Cutting stem end to appropriate size.
2. Fruit size/uniformity
3. Colour of the fruit /uniformity
4. Condition of calyx
5. Freedom from sunscald, disease spots, fruit borer, rusting and rough surface.
6. Fungicide residue in no case should exceed prescribed limits
7. T.S.S. (^{0}Brix)
8. Spoilage at the port due to diseases before dispatch to importing countries.

Chapter 14

PESTS AND DISEASES MANAGEMENT

Anar butterfly, *Deudorix isocrates* (Fab.)

The pest has been found boring into fruits of pomegranate. Besides it may attack guava, aonla, apple, ber, citrus, litchi, loquat, sapota, etc., pomegranate is the most preferred host for this pest. It may destroy up to 50 percent fruits. The female lays eggs singly on calyx of flowers or small fruits. On hatching, the caterpillars bore inside the developing fruits and are usually found feeding on pulp and seeds just below the rind; subsequently bacteria and fungi causing fruit rot also attack the infested fruits. The conspicuous symptoms of damage are offensive smell and excreta of caterpillars coming out of the entry holes, the excreta is found stuck around the holes. The affected fruits ultimately fall down and are of no use; even if the fruits be picked before falling down, these fruits have no market value.

Control : Remove and destroy all the affected fruits (fruits with exit holes). Spray Decamethrin 0.0028 percent, when more than 50 percent of fruit have set. Repeat the spray after 2 weeks with Fenvalerate 0.005 percent. In dry-rainy season Quinalphos 0.006 percent is also found effective. Remove flowering weeds especially of compositeae family in the orchard.

Fruit Sucking Moth (Othreis Fullonilca Clerck)

The moth feed nocturnally by penetrating the fruit with their highly sclerotised, heavily armoured proboscis and suck the juice from the fruit. The secondary infection of microorganisms takes place due to which fruit rots and drops down. Mote *et al.* (1991) reported 57 percent damage. The minute holes are obsered up to 8-9 per fruit. Besides pomegrante, it also attacks citrus, banana, guava, mango, papaya, lemone etc.

Pest Management

- Use of repelents like citronella oil should be tried
- Covering of fruit with paper bags, however this should be avoided where there is problem of mealy bugs.
- Use of poison baits containing malathion 20 ml along with jaggary (100 g) and vinegar (6 g) per litre of water.
- For control of semiloopers of fruit sucking moth, spray Dichlorovas or Endosulfan at 0.05 percent.
- Adjust fruit season to escape the peak activity of moths.
- Destruction of larval hosts (wild creepers).
- Use of Flood light to attract the insect
- Collect and destroy fallen rotton fruits.

Shot hole borer (*Xyleborus similes*)

This is an original pest of tea and coffee plantation in Srilanka, South India and Assam. Since 1981 onwards it has become major pest of pomegranate in Maharashtra and

Karnataka. Adult beetles of this pest are black in color. The size of female beetles is 2.0-2.5 mm long while male are almost half the size of females. Adult beetles bore holes on the lower parts of main trunk. The holes are used as nest. These holes cut through xylem and phloem, as a result yellowing starts and finally the trees may dry up. The minute pin holes on the stems are only the external visible symptoms of damage of this pest.

Fruit Sucking Moth

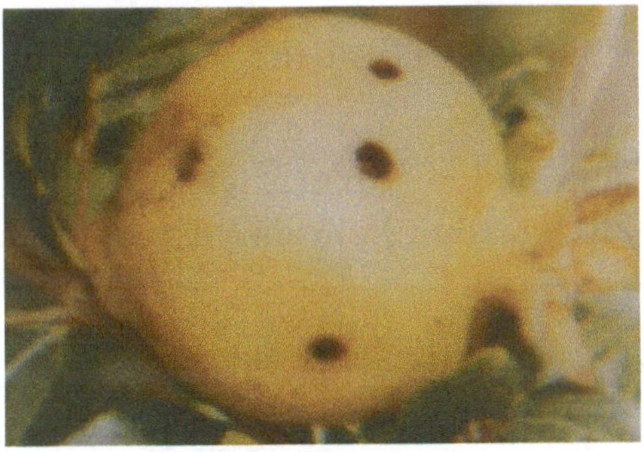

Fruit Borer

Some infested trees show heavy bearing but with reduced size of fruits. From infested tree adults migrate, within a month, to the nearest healthy trees if kept untreated and becomes a powerful source of inoculum. The rate of spread of infestation at this time will be rapid, and the whole orchard can show the symptoms in a matter of 3-6 months. From one orchard, the infestation can spread to neighboring orchards.

Pest Management

Cultural and mechanical control:

1. Keep proper distance between plants so as to avoid crowding of trees.
2. Avoid water logging and keep soil raked and aerated.
3. Infested trees should be uprooted and burnt especially the root zone.

Control : Early diagnosis with symptoms is a must. Therefore, regular visits to the orchards by growers are suggested. Signs of lateral branch yellowing to quick drying of full tree, should be immediately treated as under:

1. Drench the soil around the main trunk with a mixture of Chlorpyrifos 20 EC 2.5 ml + Monocrotophos 36 EC 1.5 ml + Tridemorph 80 EC 1 ml per liter. Use 2-3 liters of mixture/tree.
2. If pest is severe, repeat the above drenching after a month.
3. If infestation is low, drench with Azadirachtin (0.15%) @ 3 ml/liter around the main trunk @ 2-3 liters of mixture/tree.
4. Avoid water logging and keep soil racked and aerated.
5. Uproot and burn infested trees.
6. Thoroughly drench pits of uprooted trees with Chlorpyrifos 20 EC 2.5 ml/liter.
7. Drench soil with Chlorpyrifos 5 ml/liter around all uninfected trees prophylactically once in 6 months, followed by a spray on trees with Quinalphos 0.06%, followed by Azadirachtin 6 ml/liter. Avoid leaving infested trees in field after uprooting.

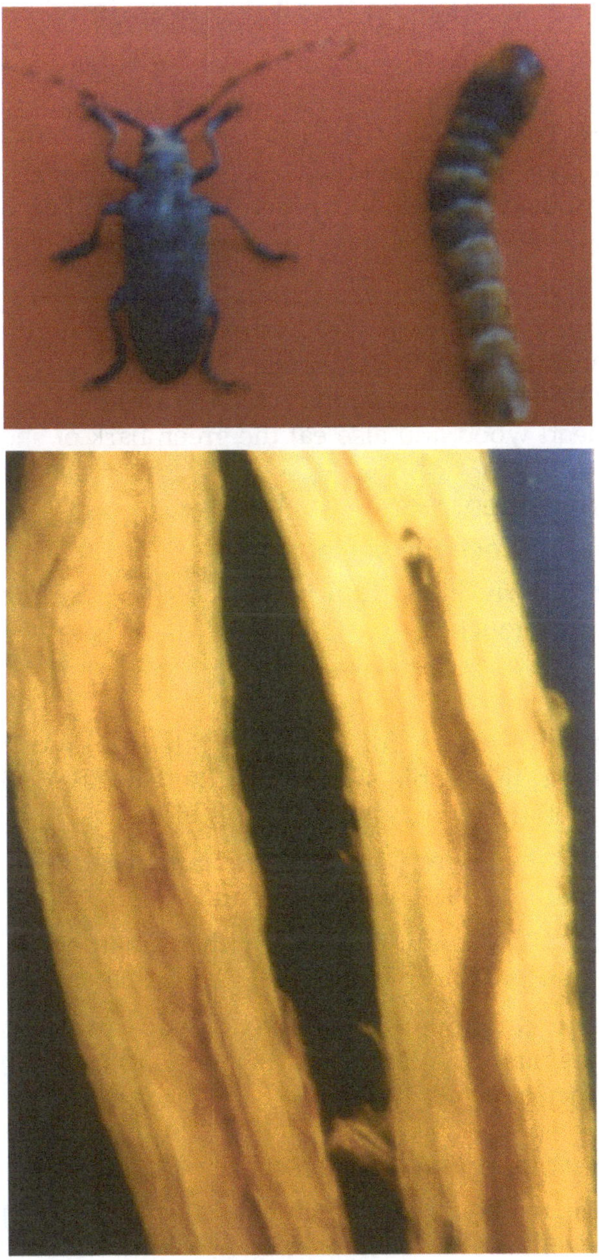

Stem Borer

Stem boring beetles (*Coelosterna spinator* Fab)

Stem boring beetles or stem borers are active during the night time. The female beetles lay eggs on the stem especially during onset of monsoon. The grub bores into the trunk; primary and secondary branches which later on leads to dieback type symptoms. Bored branch is exposed to the secondary infection, leading to slow death of tree.

The eggs hatch in fortnight and the grubs bore into the trunk, primary and secondary branches. Generally the beetles keep emerging from first summer shower in May/June, till October/November. The adults, which live for nearly 2 months, feed on dead wood and also eat the green bark of shoots.

Pest Management

Cultural and Mechanical Control

1. Detect early infestation by periodically looking out for drying branches.

Control

1. Examine branch for holes/excreta/gummosis. Inject using a disposable syringe (without needle) 5-10 ml of Dichlorvos 76 EC solution (3 ml/liter).
2. Cut off the drying portion of the branch and swab Copper-oxychloride 50% WP on cut end.
3. Spray all surrounding trees with Quinalphos 20 EC 2.5 ml/liter or Chlorpyrifos 20 EC 2.5 ml /liter.
4. It is good to give a prophylactic swab during May/June on the main trunk with mixture of Carbaryl 50 WP 8 g + Copper oxychloride 50% WP 10.

Thrips (*Rhiphiporthrips Crementatus* Hood)

It is the main species infesting pomegranate. Thrips infest leaves as well as young fruits by scrapping of the tissues and suck the cell sap. The server infestation of leaves results in leaf

Thrips Damage

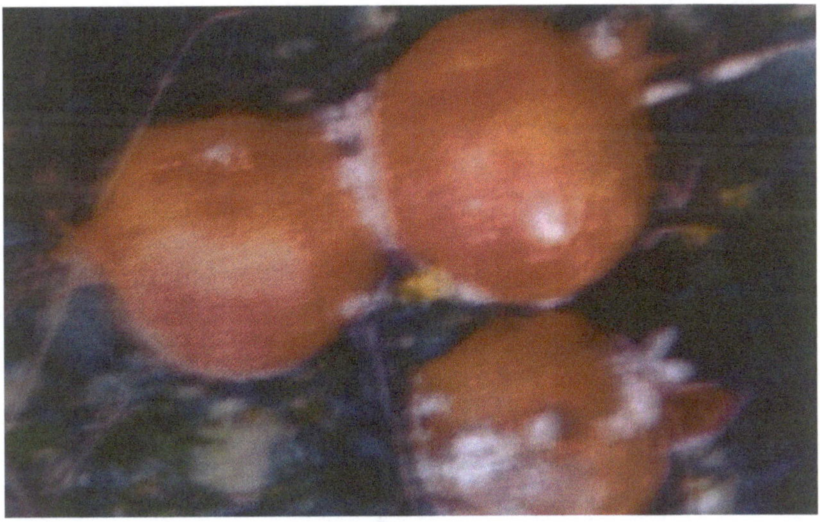

Mealy Bugs

tip curl and drying and shedding of flowers. Thrips rasp tender fruits and big fruits causing characteristics scar on the fruits and there by reducing the luster of the fruit and ultimately affects the quality of fruits for market and export.

The yield is drastically reduced. Other species of thrips recorded on pomegranate are *Retithrips syriacus* (Mayet), *Anaphothrips oligochaetus* Karny, *Ramaswamiahiella subnudula* Karny and *Scirtothrips dorsalis* Hood.Among these the first one feeds on leaves and the other three infest the flowers.

Control

Spray Dimethoate 2 ml/liter prior to flowering. If the infestation is serious spray of methyl-oxy-demeton 2 ml/liter should be repeated. Keeping the basins clean also reduces damage due to thrips. In case of species other than R.*cruentatus*, Monocrotophos 1.5 ml/liter should be sprayed.

Pomegranate aphid (Aphis punicae)

Aphids infest tender shoots, flowers, leaves and tender fruit by sucking the cell sap. The infested shoots leads to shriveling. They excrete large quantity of honeydew on which black sooty mould develops which affects photosynthesis activity adversely.

Control : Spray Dimethoate 2 ml/liter.

White fly *(Siphoninus sp.)*

Whiteflies have been observed in serious form on the leaves in pomegranate orchard where there is crowding of trees due to close planting. The nymphs and adults suck the cell sap from the lower surface of leaves and the honeydew excretion results in the development of sooty mould and ultimately affects the yield adversely.

Control : Spraying Monocrotophos 1.5 ml/liter or Acephate 1 gm/liter can control this pest.

Scale Insects (*Parasaissetia nigra* Nietn.)

Though, it is a minor pest observe in Maharashtra State in the pomegranate orchard it may sometimes appear in severe form on shoots, twigs, stem and on fruit also. The nature of damage of this pest is similar to that of mealy bugs.

Pest Management

Cultural and mechanical control

1. Remove the heavily infested shoots and burn them.

Chemical Control

1. Use of systemic insecticides for spraying.

Biological Control

1. Spraying with *Verticellum lecanii* (2 x 10^5 CFU/g) @ 3g/lit of water.
2. *Scutellista cynea* is an effective egg parasitoid of Parasaissetia nigra.

Mealy Bugs (*Ferrisia viagata Cockrell*)

Mealy bugs are found to be hard to kill, mainly because of the habitat of the pest and often they defy chemical control. Various species of mealy bugs viz., *Planococcusn citri, Nepaecoccus viridis, Maconellicoccus hirsutus and Ferrisia vinrata* are recorded on pomegranate.

Amongst the various species of mealy bugs *Ferrisia virgata Cockrell* is predominant in pomegranate orchards in Maharashtra. It has been reported throughout the tropical and sub-tropical regions of the world. It is distributed throughout Asia, Central America, Pacific Islands, West Indies and South America.

About 20 species of mealy bugs and scales have been recorded damaging pomegranate in India (Butani, 1976, Rawat et. al., 1989 and Mani and Krishnamoorthy, 1990). The nymphs and adults suck the cell sap from leaves, twigs, flowers and fruits. The damage is more pronounced on flowers and fruits. The nymphs prefer to settle at the site of attachment of fruit, inside the calyx (styler or distal end) of fruit and within the cluster of the fruits. In case of severe incidence, the flowers and fruits drop down from the tree. Similarly, the insects form thick encrustation on developed fruits, which hampers the quality of fruits. The mealy bug secretes honeydew on which the sooty mould (*Capnodium species*) develops and affects the photosynthetic activity of plant adversely. Black coating appears on fruits and hence quality gets reduced.

The incidence of mealy bugs is now a days causing serious concern in many pomegranate orchards in Maharashtra. Once the incidence is build up , it is very difficult to control this pest effectively. Similarly, the various fruiting seasons of the pomegranate respond differently to the infestation of mealy bug due to change on environmental conditions. Excess use of synthetic pyrethroids has been known to promote these pests.

Scolytid Beetle (*xyloborus fomicatus*)

Plants above the age of 5 years were mostly infested in the collar region of the plant.

Control

1. Prune the affected parts of the plant and destroy.
2. Spray Chlorpyriphos 0.06% + Dichlorvos 0.05% after pruning.
3. If on a small-scale release *Cryptolaemus montrouzieri* near the site of infestation.

Diseases

Leaf Spot

1. Leaf spot caused by *Colletorichum gloesporioides.* : Initially, minute, dull violet, black regular to irregular spots appear on leaf, which are surrounded by dark margins. Later,

these spots become bigger causing defoliation. Spray of Mancozeb (0.2%) or Kavach (0.2%) or Benlate, Bavistin or Topsin-M (0.1%) for effective control.

2. Leaf spot caused by *Phomopsis aucubicola:* Lesions were mostly marginal or sometimes scattered having buff brown color. Black *pycnidia* were unsparingly present on the surface of the leaf. Pomegranate variety Muscat white is more susceptible.

3. Leaf spot caused by *Pseudo cercospora Punicae*: The disease is marked by Light brown zonate spots. The center of the spot becomes greyish white. The infected leaves get defoliated which leads to dieback. Baycor (0.1%) is spray of found effective in controlling this disease.

Fruit Spot

1. Fruit spot caused by *Beltaraniella humicola*: Black circular spots appear on the fruit, gradually enlarge and coalesce to form big dark black spot leading to necrosis. Margin of spots varies from reddish brown in color. *In-vitro* studies have shown inhibiting effect of Miltox, Benlate (0.1 %) and copper oxychloride (0.2 %).

2. Anthracnose (*Colletotrichum gleosporioides*) is characterized as the dark brown sunken spots with dark black margins. In sever cases, these spots coalesce and form a bigger patch. This also leads to fruit cracking and fruit drop. The control measures are same as mentioned in leaf spot and treatment should start at fruit set stage.

3. Scab (*Sphaleloma punicae*): Dark brown scabby spots on the surface of the fruits. The surface becomes rough and ugly. Spray of Topsin-M (0.1%) or Baycor (0.1%) or Rovral (0.2%) at half the maturity stage of the fruit.

4. Drechsiera fruit spot caused by *Drechsiera rostsrata*: Fruits of all stages are attacked. numerous small black spots appear scattered all over the fruit. Margin of the fruit

varies from dark green to orange in advanced stages. The spots gradually enlarge end coalesce to from big dark spots of various sizes. Spraying with Diathane M-45 (2 g/l of water) or Captan (2.5 g/l of water) after the fruit formation is recommended. The subsequent 3 to 4 sprays should be given at an interval of 15 days. These sprays would also control the Alternaria fruit spots.

Fruit Rots (Post harvest diseases)

1. Phomopsis fruit rot caused by *phomopsis* sp.: Disease starts from calyx and gradually spreads over the entire fruit. One spray with Perenox can check secondary infection. The damage can be greatly reduced by 3 sprays of 0.5 % Perenox with Albolium.

2. Soft rot caused by *Rhizopus arrhizus:* Disease starts as small spot and increases in size and coalesce. The disease appears to be restricted to the rind but internally the entire fruit decays into a pulpy mess. Under dry conditions, cracking originates from the point of infection. Maximum disease has been observed at 80% humidity. Satisfactory control has been achieved by treatment of fruits with linseed, mustard or castor oil. *Cladosporium* and *Aspergillus* also causes fruit rots.

Bacterial Blight

Bacterial blight caused by *Xanthomonas campestris* pv. *Punicae*: the disease was first noticed in 1959 (Hhingorani and Singh 1959) Disease is characterized by the appearance of small irregular water soaked spots on the leaves. Spots vary from 2 to 5 mm in diameter with necrotic center of pinhead size. Spots are translucent against light. These spots turn light brown to dark brown and are surrounded by prominent water soaked margins. Numerous spots may coalese to form large patches. Severely infested leaves easily shed off. The fruit infection is also recorded. Disease also comes on stem; it stars as brown to black spots around the nodes. In advanced stages girdling and cracking of nodes occur which finally lead to breakdown of the branches. Brown to black spots appears on the pericarp with L or Y shaped cracks on the fruits. For stem infection, no suitable control measures are available at present. However, to reduce the disease, clean cultivation and orchard sanitation are recommended. The infected branches should be pasted with Bordeaux paste. Spray of copper oxychloride (0.3%) has also been recommended to control leaf and fruit infections (Hingorani and Singh, 1959).

Oily spot on fruit (*Xanthomonas axonopodis pv.punicae*)

A new disease, oily spot on leaves and a fruit was observed recently and has disastrous effect on export of fruit. The spots were circular greyish black in color with oily appearance at the center. The affected leaves then turn yellow and drop down. Two or more spots on fruits coalesce each other led to necrosis of the rind, which ultimately cracks the fruit. The infection spread from infected stem to adjoining parts.

Control

Cultural and Mechanical

Disease free planting material should be obtained.

Sanitation measure should be followed.

During rest period severe pruning of diseased and dried branches should be practiced. Apply Bordeaux paste to cut ends.

Disinfect secature with sodium hypo chlorite (1%), dettol (1%) after every cut.

Regulation of flowering to in autumn season (september-october).

Chemical

Orchard should be dusted with bleaching powder @ 20 kg/ha, copper oxide 4% @ 20 kg/ha at least 2-3 times in a year. Crop should be sprayed with streptocyclin 250 ppm, Bactronol (750 ppm) + copper oxychloride (0.25%) copper hydroxide (0.2%)/Capton (0.2%)/carbendazim (0.1%) at 15 days interval. During high humidity and rainfall dose of streptocyclin should be doubled to 500 ppm.

Use of sticker in spray solution during rainy season improved efficiency of spray.

Wilt *(Ceratocystis firmbrita.* Linn.)

Entire leaves of the plant turn yellowish. Sometimes yellowing of leaves in a single branch. Sudden wilting of leaves. Seedling wilt is also observed some of the typical symptoms are vertical stem cracking. Blue stain patches on the bark. Brown or Pink streak or patches in the xylem. The pathogen spreads through irrigation water, root contact, implements, rainwater, budding knife, secature and wind. The fungus survives in infested stem up to 190 days in the form of dormant mycelium and chlamadospores in the xylem vessels and in soil for long time.

Control

Trenching of diseased as well as adjuvent plants with carbondium (0.2%) propiconazole (0.2%) + Chlorpyoriphs (2.2%). Repeat renching after 20-30 days.

Uproot diseased plants and trench basin with carbondium (0.2 %) apply phorate (25 g/plant) or neem calce (250 g/plant) apply bioagents trichodomonas viride and pseudomonas Fluorescens.

Irrigation of the orchard with carbondium (0.2 %) + chlorpyoriphs (0.2 %) twice or thrice a year is a good preventive measure. Soil cultivation during hot summer months can also provide control of the disease.

GLOSSARY

A

Actionomorphic : Having radially symmetric shape, usually refers to the petals of a flower

Acuminate : The shape of a tip or base of a leaf or perianth segment where the part tapers gradually and usually in a concave manner.

Adnate : Joined to or attached to.

Albumen : Starchy and other nutritive material in a seed, stored as endosperm inside the embryo sac, or as periderm in the surrounding nucellar cells; any deposit of nutritive material accompanying the embryo.

Anatropous : Bent over through 180 degrees to lie alongside the stalk.

Androceium : All the male reproductive organs of a flower; the stamens

Angiosperm :	A plant producing seed enclosed in an ovary. A flowering plant.
Annual :	A plant that completes its life cycle from germination to death within one year.
Anterior :	Front; on the front side; away from the axis.
Anther :	The pollen : bearing (terminal) part of the male organs (Stamen), borne at the top of a stalk (filament).
Anthesis :	Flower bud opening; strictly, the time of expansion of a flower when pollination takes place, but often used to designate the flowering period; the act of flower bud opening.
Apex :	The tip of an organ, the growing point.
Apical :	Pertaining to the apex.
Apiculate :	Having a short point at the tip.
Aril (arillus) :	A fleshy or sometimes hairy appendage or outer covering of a seed.
Arillate :	Provided with an aril.
Auricle :	Small ear :like projections at the base of a leaf or leaf : blade or bract.
Axil :	The upper angle formed by the union of a leaf with the stem.
Axillary :	Pertaining to the organs in the axil, e.g. buds, flowers, inflorescence.
Axis :	The main or central stem of herbaceous plant or of an inflorescence.

B

Basal :	Borne on or near the base.
Bifarous :	In 2 rows; 2 :fold; pointing in directions.
Bifid :	Forked; having a deep fissure near the center.
Bilabiate :	Two : lipped.

Bipinnate : A pinnate leaf with primary leaflets themselves divided in a pinnate manner; of pinnate.

Biseriate : In two rows.

Bisexual : Having both sexes present and functional in one flower.

Blade : The flattened part of a leaf; the lamina.

Bract : A much :reduced leaf, particularly the small or scale :like leaves in a flower cluster or associated with the flowers; morphologically a foliar organ.

Bracteole : A secondary bract; a bractlet.

C

Caduceus : Falling off early, or prematurely, as the sepals in some plants.

Calyx : The outer whorl of floral envelopes composed of the sepals.

Carinate : Keeled; provided with a projecting central longitudinal line or ridge on the under surface.

Carpel : One of the flowers' female reproductive organs, comprising an ovary and a stigma, and containing one or more ovules.

Clone : A group of plants that have arisen by vegetative reproduction from a single parent, and which therefore all have identical genetic material.

Colpus (colpipl.) : Aperture within the pollen grain wall.

Concave : The interior of a curved surface.

Confluent : Merging or blending together.

Connate : United or joined as one body or organ.

Convex : Arched outward.

Cordate : Heart : shaped, often restricted to the basal portion rather to the outline of the entire organ.

Coriaceous : Of leathery texture.

Cotyledon : Seed leaf; the primary leaf or leaves in the embryo.

Crenate : Shallowly round :toothed, scalloped.

Crenulate : Finely crenate.

Cross pollination : The transfer of pollen from the anther of the flower of one plant to the flowers of a different plant.

Crustaceous : Of hard and brittle texture.

Cucullate : Finely crenate.

Cultivar : A race or variety of a plant that has been created or selected intentionally and maintained through cultivation.

Cuneate : Wedge :shaped; triangular with the narrow end at point attachment, as in the bases of leaves or petals.

Cuspidate : With an apex abruptly and sharply constricted into elongated, sharp : pointed tip.

Cyme : A broad, more or less flat :topped, determinate flower cluster, with central flowers opening first.

Cymose : Inflorescence showing the cyme arrangement.

D

Deciduous : Falling at the end of one season of growth or life, as the le of non : evergreen trees.

Decoction : Herbal preparation made by boiling a plant part in water.

Degree :day : A unit that represents one degree of difference from a given point (as 65°) in the mean daily outdoor temperature and that is used especially to measure heat requirements.

Dehiscence : The method or process of opening a seedpod or anther.

Dentate : With sharp, spreading, course indentations or teeth, perpendicular to the margin.

Denticulate : Minutely or finely dentate.

Derived : Originating from an earlier form or group.

Dichotomous : Forked, in 1 :or2 pairs.

Diocious : Having male (staminate) and female (pistillate) flowers on different plants.

Diploid : Having two sets of chromosomes.

Dipterous : Having two wings.

Distichous : Two ranked, with leaves, leaflets or flowers on opposite sides of stem and in the same plane.

Divaricate : Spreading very far apart; extremely divergent.

Downy : Covered with short and weak soft hairs.

Drupaceous : A fruit showing the characteristics of a drupe.

Drupe : A fleshy 1 :seeded indehiscent fruit with seed enclosed in a stony endocarp; stone fruit.

E

Elliptic : Oval in outline.

Emarginated : Having a shallow notch at the extremity.

Endocarp : The inner layer of the pericarp or fruit wall.

Endosperm : The starch and oil :containing tissue of many seeds.

Entomophilous : Insect pollinated.

Epigynous : Borne on or arising from the ovary; used of floral parts when the ovary is inferior and flower not perigynous.

Exalbuminous : Without albumen.

Exline : The outer coat of a pollen grain.

Exocarp : The outer layer of the pericarp or fruit wall.

F

Falcate : Scythe :shaped; curved and flat, tapering gradually.

Fascicle : A condensed or close cluster.

Faveolate : Honeycombed.

Ferruginous : Pertaining to or coloured like iron rust.

Filament : Thread; particularly the stalk of the stamen, terminated by the anther.

Filiform : Thread :shaped, long, slender and teret.

Flexuose : Zigzag; bending from side to side; wavy.

Fulvous : Dull, brownish :yellow.

Fuscous : Dusky, grayish :brown.

G

Genus : A group of related species, the taxonomic category ranking above a species and below a family.

Genotype : The genetic constitution of an organism, acquired from its parents and available for transmission to its offspring.

Glabrous : Not hairy.

Glacous : Bluish white; covered or whitened with a very fine.

Globose : Globe :shaped.

Glabrescent : Becoming glabrous with age.

Glomeruliform : Showing a compactly, clustered form.

Gynoecium : All the female parts of a flower.

H

Homonym : A scientific name given two or more times to plants of the same taxonomic rank but which are quite distinct from each other.

Hymenopterous : Having 4 membranous wings.

Hypocotyl : The axis of an embryo below the cotyledons which on seed germination develops into the radicle.

I

Indehiscent : Not regularly opening, as a seedpod or anther.

Indigenous : Native and original to the region.

Inflorescence : The flowering part of a plant and especially the mode of its arrangement.

Integuments : An outer covering or coat.

IU (international unit) : A unit used to measure the mass of certain vitamins and drugs based on their expected effects. For each substance to which this unit applies, there is an international agreement specifying the biological effect expected with a dose of 1 IU. Other quantities of the substance are then expressed as multiples of this standard Example: 1 IU represents 45.5 micrograms of a standard preparation of insulin or 0.6 microgram of a standard preparation of penicillin. Consumers most often see IU's on the labels of vitamin packages: the equivalent of 1 IU is 0.3 microgram (0.0003 mg) for vitamin A, 50 micrograms (.05mg) for vitamin C, 25 nanograms (0.000 025 mg) for vitamin D 2/3 milligram for vitamin E.

L

Lateral : Side shoot, bud etc.

Lamellae : A thin, flat plate or laterally flattened ridge.

Lanceolate : Shaped like a lance head, several times longer than wide, broadest above the base and narrowed toward the apex.

Lenticellate : Having a body of cells as a pore, formed on the periderm of a stem, and appearing on the surface of the plant as lens :shaped spot.

Locular : Having a cavity or chamber inside the ovary, anther or fruit.

M

Membranous : Thin in texture, soft and pliable.

Mesocarp : The fleshy middle portion of the wall of a succulent fruit between the skin and stony layer.

Monophyletic : Descended from a single ancestral line, see also: polyphyletic.

Mucronate : Terminated abruptly by a distinct and obvious spur or spiny tip.

N

Naturalized : To cause a plant to become established and grow undisturbed as if native.

Nectar : Sweet secretion of glands in many kinds of flower.

Necrotic : Death of cells or tissues through injury or disease.

Nectiferous : Producing nectar.

Nervules : Small/minute nerves or veins.

O

Obconical : Inversely conical, having the attachment at the apex.

Oblique : Slanting, unequal sided.

Obovate : Inverted ovate; egg :shaped, with the broadest part above.

Obtuse : Blunt or rounded at the end.

Octaploids : Having 8 times the basic number of chromosomes.

Orbicular : Circular.

Osseous : Bony.

Ovary inferior : With the flower :parts growing from above the ovary.

Ovary superior : With the flower : parts growing from below the ovary.

Ovate : Egg :shaped, with the broader end at the base.

Ovule : The body which after fertilization becomes the seed.

P

Panicle : A loose irregularly compound inflorescence with pedicel late flowers.

Paniculate : Borne in a panicle.

Papillose : Bearing minute, nipple :shaped projections.

Paripinnate : A pinnate (compound) leaf with all leaflets in pairs.

Pedicel : A tiny stalk; the support of a single flower.

Pendulous : More or less hanging or declined.

Perianth : The floral envelope consisting of the calyx and corolla.

Pericycle : The tissue of stele lying just inside the endodermis.

Perigynous : Adnate to the perianth, and therefore around the ovary and not at its base.

Petal : A division of the corolla; one of a circle of modifies leave immediately outside the reproductive organs, usually brightly colored.

Petiole : The stalk of a leaf that attaches it to the stem.

Phenotype : The morphological, physiological, behavioral, and other outwardly recognizable forms of an organism that develop through the interaction of genes and environment.

Pilose : Hairy, especially with soft hairs.

Pilosulous : Minutely pilose.

Pinnate : A compound leaf consisting of several leaflets arranged on each side of a common petiole.

Polygamous : Bearing male and female flowers on the same plant.

Polyphyletic : Having members that originated, independently, from more than one evolutionary line.

Polyploidy : Having more than two sets of chromosomes.

Polyporate : Pollen grain with many apertures.

Porate (see tricoporate, polycolporate) : Describes a pollen grain, which has rounded apertures only.

Prolate : Having flattened sides due to lengthwise elongation.

Propagate : To produce new plants, either by vegetative means involving the rooting or grafting of pieces of a plant, or sexually by sowing seeds.

Protandrous : Refers to a flower where the shedding of the pollen occurs before the stigma is receptive.

Psilate : Referring to a pollen grain having a smooth surface.

Pubescent : Covered with hairs, especially short, sort and down -like.

Pulvinus : A swelling at the base of a leaf or leaflet.

Putamen : The shell of a nut, the bony part of a stone fruit.

Pyriform : Pear shaped.

R

Raceme : A simple inflorescence of pediceled flowers upon a common more or less elongated axis.

Rachis : The main stalk of a flower cluster or the main leafstalk of a compound leaf.

Ramification : Branching

Reticulate : In the form of a network, net veined.

Retuse : With a shallow notch at a rounded apex.

Rootstock : The root system and lower portion of a woody plant to which a graft of a more desirable plant is attached.

Rotundate : Nearly circular; orbicular to oblong.

Rufous : Reddish :brown.

Rugose : Wrinkled.

Rugulose : Covered with minute wrinkles.

S

Scandent : Climbing but not self :supporting.

Scarify : To scar or nick the seed coat to enhance germination.

Sclerenchymatous : Tissue composed of cells with thickened and hardened walls.

Scurfy : Covered with tiny, broad scales.

Self pollination : The transfer of pollen from the anther of a flower to the stigma of the same flower, or difference flowers on the same plant.

Sepal : A division of calyx; one of the outermost circles of modified leaves surrounding the reproductive organs of the flower.

Sericeous : Bearing fine, usually straight, appressed hairs.

Serrate : Having sharp teeth pointing forward.

Serrulate : Finely serrate.

Sessile : Without a stalk.

Sheath : A tubular envelop.

Spathulate : Gradually narrowing downward to a summit; spoon shaped.

Spine scent : 1. Having spines, 2.terminating in a spine, 3. Modified to form a spine.

Stamen : One of the male pollen :bearing organs of the flower.

Staminode : A sterile stamen, or any structure without anther corresponding to a stamen.

Stigma : That part of a pistil through which fertilization by the pollen is effected.

Stipule : An appendage at the base of a petiole, often appearing in pairs, one on each side, as in roses.

Strigose : Beset with appressed, sharp and stiff hairs.

Style : The usually attenuated portion of the pistil connecting the stigma and ovary.

Subulate : Oswl shaped.

T

Tetraploid : Having four sets of chromosomes (twice the normal number chromosomes).

Testa : The outer seed coat.

Tomentose : Covered with a thick felt of radicals; densely pubescent with matted wool.

Tomentulose : Rather tomentose.

Tomentum : Closely matted, wooly hairs.

Trabecular : A transverse partition, complete or incomplete.

Transverse : Cross :wise in position.

Tricoporate : Having three apertures in the pollen grain wall.

Trifid : Deeply divided or left in three parts.

Tropism : The movement of an organism in response to an external source of stimulus, usually toward or away from it.

Truncate : Ending abruptly, as if cut off transversely.

Tuberculate : Bearing tubercles, covered with warty lumps.

U

Unguiculate : Narrowed, clawed.

V

Valvate : Open by valves.

Z

Zygomorphic : Capable of division by only one plane symmetry.

BIBLIOGRAPHY

Akoi, S. Arakik kango, K. and Katayama, O, (1970) Phenylanine ammonia lyase activities in Japanese chestnut, strawberry, apple fruit and bracken. Journal of Food Science and Technology 17: 507-511.

Anonymars (1979) Annual report, Dept. of Horticulture Mahatma Phule Krishi Vidyapeeth, Rahuri, Maharashtra.

Anonymous (1996) Annual Report, Indian Institute of Horticultural Research, Bangalore.

Anonymous (1996) Coordinated Fruit Research in Indian Arid Zone : A Two Decade Profile. pp. 36.

Artes, F, Gains Martin, J. and Martinez J.A., (1996) Controlled atmospheric storage of pomegranate. J. Zeitschrift. Fuer-Lensmittlel-untersuchung-und-Forschung, 203(1) 33.

Artes, F, Tudala, J.A. and Villaescusar (2000) Thermal pH treatments for improving pomegranate quality and shelf life J.P.H. Biol. Technol.18 (3), 245

Bacharaj S.M. and Bhosale V. (2005) Use of liquid nutrient fertlizers through fertigation in pomegranate.

Bankar, G.J.and Prasad, R.N. (1992) Performance of important pomegranate cultivars in arid regions. Annals of Arid Zone 31(3): 181-3.

Ben-Arie R, Segal N, and Guelfat Reich S. (1986) The Maturation and ripening of the Wonderful pomegranate, J.Amer. Soc. Hort.Sci.109 (6), 898.

Bhandari, M.M. (1979), Practicals in Plant Breeding Oxford & IBH Publishers Co. New Delhi. PP. 23,52,79.

Bhargava, B.S. and Dhandhar D.G. (1987) Leaf sampling techniques for pomegranate (*Punica granatum* L.) Prog. Hort. 19(3-4).

Bhojappa, A.P. and Singh, R.N. (1975) Feeder root distribution pattern of young and old mango (*Mangifera indica* L.) trees. Indian Journal of Horticulture 32:123-127.

Bilsil, A. and Cevic, I. (1997) Studies on stability of some pomegranate varieties for freezing preservation publication II-Kontrol-Laboratuvar-Mudurlugu-Canakkal No. 37, 12pp.

Bohlmann, T.E.1962 Why does a fruit crack. Farming Biology of Africa. 36 (7): 12-13.

Burton, G.W. De Van .W. 1953 Estimating heritability in tall fescue (*Festuca aruninancea*) From replicated clonal material. Agronomy Journal 45:478-81.

Chaturvedi, K.N., Rai, J.N. and Gupta D.B. (1980) Variability in grape. Ind J. of Hort. 37:262-4

Du, C.T., Wang, P.L. and Francis F.J. (1975) Anthocyanins of pomegranate Journal Food Science, 40: 417-418.

Edward Raja, M (2008) Identification of important micronutrient disorders in pomegranate in Maharashtra Dalimb Vritt Smaranika Pp: 149-151.

Firke, N.N. and D. B. Kumbhar (2002) J. Maharashtra Agric. Univ., 27(2): 146-148 Mahatma Phule Krushi Vidyapeeth, Rahuri-413 722

Gafizov, G.K., Semochkina, L.G. and Gakharmonov, M.S. (1990) Method for improving the color stability of pasteurized natural pomegranate juice during storage. USSR Patent SU 1 535 517.

Guar, A.C., Neelkanthan, S., Dargan, K.S. (1971) Organic manures, ICAR.Pub. pp : 3-10

Hayes, W.B. (1953) Fruit growing in India 3rd Rev. Ed., Kitabisthan, Allahabad, India.

Hingorani, M.K. and Singh, N.J. (1959) *Bathomona punicae* sp. on Punica granatum L. Indian J. Agric. Sci., 29:45-48.

Hiwale S.S. (2001) Evaluation of pomegranate germplasm, Annual report, CIAH Bikaner,2001.

Hiwale S.S. (2002) Root distribution pattern in Ber (*Zizyphus Mauritania* Lamk). Annals of Arid Zone 41(1) 89-91.

Hiwale S.S., Bagle, B.G. and More, T.A. (2009) Root distribution pattern in pomegranate cv. Ganesh. Paper presents in paster session of International symposium on pomegranate including mediterranian fruits held at UAS Dharwad from 23rd June to 27th June.

Hiwale, S.S. (2003) Effect of Organic and inorganic fertilizers on growth, Yield and quality of pomegranate cv. Ganesh. Paper presented in "National Seminar on Organic Farming in Horticulture for Sustainable Production on 29-30 Aug.2003.

Hiwale, S.S. (2005) Effect of fruit thinning on growth, Yield and quality of pomegranate cv. Ganesh. Paper presented in First Hort. Congress Organized by Hort. Society of India at Pusa, New Delhi 6-9 Nov. 2005.

Hiwale, S.S., Dhandhar, D.G. and Bagle, B.G. (2006) Rejuvenation of pomegranate through non-selective pruning. Paper presented in "National Symposium on Production, Utilization and Export of Underutilized Fruits with Commercial Potentialities" Held at BCKV, Kalyani, West Bengal on 22-24 November 2006.

Hiwale, S.S., Dhandhar, D.G., And Bagle, B.G. (2008) Effect of Biofertilizer on Growth, Yield and Quality of Pomegranate cv. Ganesh. Haryana J of Hort. Science. (Accepted).

Hussein, M.L.; Devikar, G.Y; Singh, G and Jaiswal (2004) Quality evaluation of pomegranate seed and powder. *Spice India, July 2004,* 17: 28-31.

Hyodo, H. (1971) Phenylalinine ammonia lyase. In strawberry fruits Plant Cell Physiology, 12: 989-991.

Iboyaima Singh, Madaiah, N and Nanjundaswamy, A.M. (1993) Preliminary studies on clarification of fruit juices by Ultra filtration. Ind. Food. Packer, 47 (5): 9.

Jagannath, S. Raju, B, Krishna Manohar, R., Narayan Gowda, J.V. and Khan M.M. (1999) Yield and yield attributes of pomegranate cv. Jyothi as influenced by Drip Irrigation and Black Polyethylene mulch. Paper presented in National Seminar on Problems and Prospects of Micro Irrigation - A Critical Appraisal held at GKVK, campus, Bangalore pp 46-51.

Jalicop, S.H., Tiwari, R.B., Sampath Kumar. (2002) Amlidana: a new pomegranate hybrid. Indian Hort. July Sept. 2002 PP.

Josan, J.S., Jawanda, J.S.and D.K. Uppal (1979) Studies on floral biology of pomegranate III mode of pollination, fruit development and fruit cracking. Punjab Hort. J.:136-138.

Khodade, M.S. (1987) Studies on physico- chemical change during growth and development of pomegranate fruits (*Punica granatum* L.) M.Sc. (Agri) thesis submitted to MPKV. Rahuri.

Khrameeva, N.P. Kosalkina O.T. and Ruanet V.V. (1987) Spectral characteristics of pomegranate juice after treatment with stabilizers .J. Izavenstiya-Vysshikh-Uchebnykh-Zavednil-Pishchevaya. teknologeya No.3: 96.

Kulkarni, A.P., Aradhya, S.M. and Divakar, S. (2005) Chemical changers and antioxidant activity in pomegranate arils during development. Food.Chem.93: 319.

Liberman, R. G. (1993) Effect of biocontrol agents on plant growth. Cambodia, Proceedings of International plant propagation society 43:249-252.

Mahajan, B.V.C.; Chopra, S.K.: Sharma, R.C. (1992) Processing of Wild Pomegranate (Punica granatum L.) for Anardana: Effect of Thermal Treatments and Drying Modes on Quality. *J. Food Science and Tech., 1992 Sept/Oct;* 29: 327-328.

Meena, K.K. Room Singh and Singh S. K. (2003) Genetic variability and genetic advance to average fruit weight and its components traits in pomegranate (*Punica granatum*) genotypes. Indian J. Agril. Sciences73 (11) 630-2.

Misra, A.R. and Khan, I. 1981 Trichlorophenoxy acetic acid and micronutrient on fruit size, cracking and maturity in litchi cv. Rose scented. Progressive Hort, 13(3-4): 87-90.

Murukute, A., Patil S. Patil B.N. & Mayakumari (2002). South Indian Hort., 50 (1-3): 49-55.

Palmenac, M (1971) A contribution to studies on fruiting of pomegranate varieties in the Be district. Jugoslovenko Voscsatro, 5 (17/18): 233-240. H.A. 43: 4030.

Pandey, G. and Bist H.S. (1998) Variability, correlation and path analysis in pomegranate germplasm. The Hort. J., 11:7-12.

Panse, V.G. and Sukhatme, P.V. (1978) Statistical methods for Agril. Workers, pp 381. ICAR, New Delhi.

Patil, P .B., Patil, C .P., Swamy, G.S.K. and Kumar, N. (2004). Influence of micro propagation on rooting of pomegranate cuttings. In Proceedings organic farming in Horticulture, pp: 209-211.

Pawar, S.K. 1993 Pruning studies in (*Punica granatum* L.) cv. Ganesh M.Sc. (Agri) thesis submitted to MPKV. Rahuri.

Phadnis, N.A. (1974) Pomegranate For desert and juice Indian Hort.19 (3): 9-13.

Prabhu-Desai, V.G. 1989 Investigation on internal breakdown of pomegranate fruits (*Punica granatum* L.) PhD thesis submitted to MPKV. Rahuri.

Prasad R.N.and G.J. Bankar, (2000) Evaluation of pomegranate cultivars under arid conditions. Indian J. Hort. 57(4) 305-308.

Prasad, R. N. and Bankar, G.J. (2003) Pomegranate cultivation. Indian Horticultural Journal, pp 26-29. Jan- March 2003.

Prasanna Kumar, B. and Dhandar D.G. (1996) Effect of NPK on yield and quality of pomegranate cv. Bassein seedlers Indian J. of Hort. (53) : 114-117.

Pruthi, J.S. (2001) Anardana. In: Minor spices and condiments-Crop management and post harvest technology, ICAR, Pusa, New Delhi. Pp 393-405.

Purohit, A. G. (1985) Soft seeded ness of commercial pomegranate varieties. Indian J. of Agril. Sciences 55(5): 367-368.

Purohit, A.G. (1983) Evaluation & pomegranate cultivars for growth, yield quality and cracking paper presented in 2nd workshop on Arid Zone fruit research.

Rage, A.R.and Pai, J.S. (1999) Development of thermal process for clarified juice of pomegranate (*Punica granatum*). J.Food.Sci. Technol.India. 36(3): 261.

Randhawa, G.S.Singh, J.P. and Malik, R.S. 1958 Fruit cracking in some fruit trees with special reference to lemon (*Citrus limon*). Indian J. of Hort., 15:6-9.

Raturi, G.B. and S.S.Hiwale (1991) Regulation of cropping in pomegranate under rainfed conditions. Indian J. of Horticulture 48(1): 13-15.

Riaz, R.A. and Elahi, M. (1992) Studies on preparation and storage stability of Carbonated pomegranate drink J.Sci. Ind.Res. (Pakistan) 35 (11): 467.

Riker, A. J., Benefield, U. M. Wright, W.H., Ketit, G.W. and Sagen, H. E., (1930). Studies on infectious hairy root of nursery apple trees. Journal of Agricultural Research 41:507-540.

Sawant, B.A. 1993 Biochemical investigations of pomegranate (Punica granatum L.) arils in relation to blackening. M.Sc. (Agri) thesis submitted to MPKV. Rahuri.

Sharma, S.D. and Sharma, V.K. (1990) Variation for chemical characters in some promising strains of wild pomegranate (*Punica granatum* L) J. Euphytica 49(2): 131.

Sheikh, M.K. and Patil, S.R (2005) Pomegranate Smaranica pp: 70. Published by All India pomegranate growers research association Pune.

Shete, M.B. 1998 Effect of varieties, stage of harvest and bahar treatments on internal breakdown of pomegranate fruits (*Punica granatum* L.) PhD thesis submitted to MPKV. Rahuri.

Shinde, D.G. (1977) Effect of N, P and K on growth, Yield and quality of pomegranate (*Punica granatum* L.) M.Sc. (Agri) Thesis M.P.K.U.Rahuri.

Sonwane, P.C. and Desai, U.T. (1989) Performance of staggered cropping in Pomegranate. J. Maharashtra Agric. Univ. 14(3): 341-342.

Sonwane, P.C. and Desai, U.T. (1989) Performance of staggered cropping in Pomegranate. J. Maharashtra Agric. Univ. 14(3): 341-342.

Waskar, D.P., Khedkar R.M. and Garnade, V.K. (1999) Effect of post harvest treatments on shelf life and quality of pomegranate in evaporative cool chamber and ambient conditions. J. Food. Sci. Technol., India 36 (2), 114

Waskar, D.P., Khedkar, P.B. and Garnade, V.K. (1999) Effect of post harvest treatments on storage behavior of pomegranate fruits under room temperature and cold storage. Ind. Food. Packer, 53 (2), 5.

Annexures

Annexure - I

Cost of cultivation - pomegranate (400 plants/ha)

			I year	II year	III year
1.	Preparation of land	-	1500	—	
2.	Digging and filling of pits (Rs. 10 / pit)	-	4000	—	
3.	Plant material (Rs. 15/plant	-	6000	-	
4.	Planting and Staking (Rs. 2/plant)	-	800		
5.	Tool / Implements	-	5000		
6.	Manures and fertilizers	-	3000	4500	4500
7.	Plant protection	-	1500	2000	2,000
8.	Inter culture	-	2000	1500	1500
9.	Harvesting	-	—	—	500
10.	Miscellaneous	-	1,000	2,000	2,000
			20800	10000	10500

Annexure - II

List of Pesticides to be analyzed

S.	Chemicals	EU Value in mg/kg	UK	Nether-lands	Ger-many	Codex
	A) Organochlorine					
1	Aldrin (expressed as dieldrin)	0.01	0.01	0.01	0.01	0.05
2	Chlordane X (cis & trans)	0.01	0.01	0.01	0.01	
3	Chlorothalonil	0.01	0.01	0.01	0.01	0.05
4	DDT (all isomers)	0.05	0.05	0.05	0.01	
5	Dicofol	0.02	0.02	0.02	0.02	5.00
6	Dieldrin (See Aldrin)	0.01	0.01	0.01	0.01	
7	Endosulphan (All isomers)	0.05	0.05	0.05	0.05	1.00
8	Endrin	0.01	0.01	0.01	0.01	
9	HCH (alpha & beta)	0.01	0.01	0.01**	0.01	
10	Heptachlor	0.01	0.01	0.01	0.01	
11	Lindane	0.01	0.01	0.01	0.01	0.50
	B) Organophosphorus					
12	4-bromo-2chlorophenol	0.01	0.01	0.01	0.01	
13	Acephate	0.02	0.02	0.02	0.02	
14	Chlorfenvinphos	0.02	0.02	0.05	0.05	
15	Chlorpyriphos	0.05	0.05	0.50	0.50	1.00
16	Chlorphyriphos-methy1	0.05	0.05	0.20	0.20	0.20
17	Diazinon	0.02	0.02	0.02	0.02	
18	Dichlorvos	0.01	0.01	0.01	0.01	
19	Dimethoate(including Omethoate)	0.02	0.02	0.02	0.02	1.00
20	Ethion	0.01	0.01	0.50	0.50	
21	Etrimphos	#	#	0.05	0.02	
22	Fenitrothion	0.01	0.01	0.50	0.50	0.50
23	Fosety1-A1	#	#	0.02	5.00	
24	Iprobenphos	#	#	#	0.01	8.00
25	Malathion	0.50	0.50	0.50	0.50	
26	Methamidophos	0.01	0.01	0.01	0.01	
27	Monocrotophos	#	#	0.05	0.01	
28	Omethoate (refer to Dimethoate)	-	-	-	-	
29	Oxydemeton-methy1	0.02	0.02	0.02	0.02	
30	Parathion ethy1	0.05	0.05	0.05	0.05	
31	Parathion-methy1	0.02	0.02	0.02	0.02	

Contd...

32	Phorate	0.05	0.05	0.05	0.05	
33	Phosalone	1.00	1.00	1.00	1.00	2.00
34	Phosphaamidon	0.01	0.01	0.20	0.15	
35	Profenophos	0.05	0.05	0.05	0.05	
36	Quinalphos	0.05	0.05	0.05	0.05	
37	Traizophos	0.01	0.01	0.02	0.02	0.20
	C) Synthetic Pyrethorids					
38	Cyfluthrin	0.02	0.02	0.30	0.30	
39	Cypermethrin	0.05	0.05	0.50	0.50	2.00
40	Deltamethrin	0.05	0.05	0.20	0.20	0.05
41	Ethofenprox (Ethofenprox)	#	#	0.01	1.00	1.00
42	Fenvalerate & Esfenvalerate (sum of RR & SS isomers)	0.02	0.02	0.1	0.10	2.00
43	Fenvalerate & Esfenvalerate (sum of RS & SR isomers)	0.02	0.02	0.02	0.10	
44	Lambda-cyhalothrin	0.02	0.02	0.20	0.20	0.2
45	Permethrin	0.05	0.05	0.05	0.05	2.00
	D) Triazines					
46	Atrazine	0.05	0.05	0.10	0.10	
47	Simazine	#	#	0.1	0.10	
	E) Acylamino acid fungicides					
48	Metalaxy1	0.05	0.05	2.00	2.00	1.00
	F) Carbamates					
49	Carbary1	0.05	0.05	2.0	2.00	1.00
50	Carbofuran	0.02	0.05	3.00	3.00	5.00
51	Carbosulfan	0.05	0.05	0.05	0.05	
52	Indoxacarb	0.02	0.02	0.10	0.30	
53	Methomy1	0.05	0.05	0.05	0.05	2.00
54	Thiodicarb (see Methomy1)					
	G) Pyrimidines					
55	Fenarimo 1	0.02	0.02	0.30	0.30	0.30
	H) Triazoles					
56	Bitertanol	0.05	0.05	0.05	0.05	2.00
57	Flusilazole	#	#	0.05	0.01	
58	Hexaconazole	0.02	0.02	0.10	0.10	0.10
59	Myclobutanil	0.02	0.02	1.00	1.00	0.50

Contd...

60	Penconazoale	0.05	0.05	0.20	0.20	0.20
61	Propiconazole	0.05	0.05	0.50	0.50	0.50
62	Tebuconazole	#	#	2.00	2.00	0.50
63	Triadimefon	0.1	0.1	2.00	2.00	0.50
64	Triadimenol	0.1	0.1	2.00	2.00	0.50
65	Difenoconazole	#	#	0.05	0.05	
	I) Dicarboximides					
66	Captafol	0.02	0.02	0.02	0.02	
67	Captan	0.02	0.02	3.00	3.0	
68	Iprodione	0.02	0.02	10.00	10.00	5.00
	J) Benzimidazole					
69	Carbendazim (Including Benomy1)	0.1	0.1	0.3	0.30	3.00
70	Thiophanate-methy1	0.1	0.1	0.1	0.10	3.00
	K) Dithicocarbamates					
71	Carbon disulfide (Mancozeb, Maneb, Propineb, Metiram, Thiram and Ziram collectively estimated as CS_2 Thiram	0.05	0.05	3.00	2.00	4.00
72	Ethylene thiourea	#	#	0.02	0.05	
	L) Nicotinoids					
73	Acetamiprid	0.01	0.01	#	0.01	
74	Clothianidin	#	#	#	0.50	
75	Imidacloprid	#	#	1.00	1.00	
76	Thiacloprid	#	0.02	0.02	0.01	
77	Thiamethoxam	#	#	0.05	0.01	
	M) Dinitrophenol					
78	Dinocap	#	#	0.05	0.50	
	N) Alliphatic Nitrogen fungicides					
79	Cymoxanil	#	#	0.05	2.00	
	O) Morpholine					
80	Dimethomorph	#	#	0.05	2.00	
81	Tridemorph	0.05	#	#	#	
	P) Natural Product Derivative					
82	Buprofezin	#	#	0.05	0.02	
83	Cartaphydrochloride	#	#	1.00	0.01	

Contd...

84	Emamectin Benzoate	#	#	#	0.01
85	Spinosad (sum of Spinosyn A+D)	0.50	0.50	0.20	0.20
86	Abamectin	0.01	0.01	0.01	0.01
	Q) Substituted Thiourea				
87	Difenthiuron	#	#	#	0.01
	R) Others				
88	Azoxystrobin	0.05	0.05	2.00	2.00
89	Fipronil	#	#	0.01	0.01
90	Flufenoxuron	#	#	0.05	0.10
91	Oxyfluorfen	#	#	#	0.01
92	Propargite	#	#	10.00	3.00
93	Fenpyroximate	#	#	#	0.50

No MRL exists and therefore the chemicals should not be detected in the sample. If detected above 0.01mg/kg then the sample will be treated as failed.

(*Source:* EC Regulations, updated up to December 2007).

Index

B

D

E

F

N

O

P

T

U

FSC
www.fsc.org
MIX
Papier | Fördert gute Waldnutzung
FSC® C083411

Zeitfracht Medien GmbH
Ferdinand-Jühlke-Straße 7
99095 Erfurt, Deutschland
produktsicherheit@kolibri360.de